# Idea Engineering

# Idea Engineering

## Creative Thinking and Innovation

La Verne Abe Harris

MOMENTUM PRESS, LLC, NEW YORK

*Idea Engineering: Creative Thinking and Innovation*
Copyright © Momentum Press®, LLC, 2014.

All rights reserved. No part of this publication may be reproduced, stored in a retrieval system, or transmitted in any form or by any means—electronic, mechanical, photocopy, recording, or any other—except for brief quotations, not to exceed 400 words, without the prior permission of the publisher.

First published by Momentum Press®, LLC
222 East 46th Street, New York, NY 10017
www.momentumpress.net

ISBN-13: 978-1-60650-472-7 (hardback, case bound)
ISBN-13: 978-1-60650-473-4 (e-book)

Momentum Press Industrial, Systems, and Innovation Engineering Collection

Collection ISSN: Forthcoming (print)
Collection ISSN: Forthcoming (electronic)

DOI: 10.5643/9781606504734

Cover design by Jonathan Pennell
Interior design by Exeter Premedia Services Private Ltd., Chennai, India

10 9 8 7 6 5 4 3 2 1

Printed in the United States of America

This book is dedicated to my IDEA Gurus at Arizona State University and Purdue University as well as my grandchildren: Kasey, Jacob, Eric, Shane, Cody, Karissa, Dylan, Hope, Austin, Ian, Kenzie, Aiden, and Grace.

# Abstract

Engineers and technologists often operate from a worldview of "ones and zeros." My mission in this book is to interject the colorful world of creative thinking to help engineers and technologists learn to think and work differently. This "idea engineering" becomes the driving force, transforming engineers and technologists into creative thinkers and innovators. Throughout the book, case studies and first-hand experience anecdotes can be found.

The material in this book is organized to take the reader through basic concepts and techniques of creative thinking and innovation leading to the solving of engineering and technological challenges. It provides an overall understanding of how idea engineering can transform an individual and a company to formulate and apply the best possibilities.

The target audience is university-level students and practitioners. Upper division undergraduates and graduate students in engineering and technology (i.e., engineering education, industrial engineering, engineering technology, mechanical engineering, computer graphics technology, computer science, etc.) will benefit from the content of this book, as will practitioners from an engineering, technology, or science background. This book is also written for engineering and technology practitioners as a part of a training collection for libraries and professional organizations. In addition, the material in this book can be applied to coursework in business, communication, management, and applied creative arts. It also can serve as a good overview for the general public interested in the creative thinking and innovation subject matter.

The book content can be used as the foundation for a one-credit course or a part of a three-credit capstone design course, which incorporates creative thinking and innovation into the coursework or seminar.

# Keywords

creative thinking, innovation, ideas, idea generation, idea engineering, idea teams, visionaries, innovators, entrepreneurs, problem solving, inventors, brainstorming, creativity, business, technologists, engineers, political, economic, and socio-constructed innovations, resistance to change, innovative culture, obstacles

# Contents

*Foreword* ................................................................... xi

*Love & Gratitude (Acknowledgments)* ........................... xv

Chapter 1    Getting Your Mind Right ................................ 1

Chapter 2    Visionaries and Innovators ............................ 17

Chapter 3    Idea Engineering Overview ........................... 35

Chapter 4    Organizational Roadblocks ........................... 43

Chapter 5    Models of Innovation ................................... 61

Chapter 6    Identifying the Problem ............................... 83

Chapter 7    Creative Thinking ........................................ 95

Chapter 8    Selecting the Best Idea .............................. 121

Chapter 9    Designing and Testing an Idea ..................... 133

Chapter 10   Worldviews .............................................. 157

Chapter 11   Final Thoughts ......................................... 169

*About the Author* ...................................................... 175

*Notes* ....................................................................... 179

*Bibliography* ............................................................. 191

*Index* ....................................................................... 199

# Foreword

*There is no passion to be found in settling for a life that is less than the one you are capable of living.*
—Nelson Mandela

When I was approached to write a book on creative thinking and innovation, I only had one request. I wanted to have a personal and intelligent conversation with the reader that they could apply to their career, as well as their life. I did not want to write a theoretical academic exposé devoid of anecdotes and real world experience. I wanted to provide readers with insights into bringing creative thinking and innovation into any business, job, or university. The publisher granted my wish. So now our conversation begins....

Everyone can learn to be a creative thinker. Everyone can improve his or her skills in innovative problem solving. But first you must make a promise to yourself. The same one I made to myself. I made a promise that I would never forget what it was like to be a child. Just remember the sense of wonder you had as a child. Children are creative, curious, innovative, and open to possibilities. Children ask questions, and live in the here and now. They are not afraid to take risks. Sadly, society removes the childlike nature out of most of us, and fear takes over. I hope that this book will begin to replace that fear with the creative thinking of a designer, inventor, and entrepreneur.

The material in this book is organized to take the reader through basic concepts and techniques of creative thinking and innovation leading to the solving of engineering and technological challenges. It provides an overall understanding of how idea engineering can transform an individual and a company to formulate and apply the best possibilities.

When I began my career in the private sector as a graphic designer and illustrator, I always seemed to notice that there were missing pieces of the puzzle, and so many thoughts flew through my head. "There has got to be a better and more efficient way of doing this." "Just because it has always been done this way, is not an answer." "Why are we not exploring the use of machines to do a portion of this work, so that we can spend more time on the creative part?" This is usually out of the realm of the job of a graphic designer, but I was always curious.

I remember many years ago during the transition from analog to digital, I identified production problems, which were caused by lack of knowledge and training in another department. It affected my ability to get my part of the job through the digital pipeline. I spent a good part of the night on the phone troubleshooting with the untrained and frustrated employee, so that we could meet deadline. None of his managers were on duty at the time. All he knew to do was to push a red button. He did not know why or what to do if it did not work, and unfortunately, it did not work. At the time I was not in management and the problem was not within the scope of my responsibilities. I could have said, "Well, I did my part!" and left, but I did not. His supervisor was called in at 3:30 in the morning to help us meet deadline. I documented my troubleshooting steps by writing an instructional sheet on how to send documents digitally through the production pipeline, along with emergency phone contacts for both departments. After that episode, the employees in that department were put through a complete training workshop.

Wherever there was missing knowledge, I looked to that as an opportunity for me to learn something new. I began to observe and study why and how a process was done and where the cog was in the pipeline. In addition to my job, I helped with the missing piece. Soon people were coming to me for consultation, since I now had the knowledge. It was not long before I had more opportunities.

If this makes any sense at all, know that this is how I combated boredom. For once I learned a job and was competent at it, I wanted to move on to something else. I was able to have six different positions in the same company doing this, and it was a joy to me. In the academic world, being a professor at two major research universities and heading a research and development laboratory enabled me to have the same job, but get the same satisfaction of continuing to apply my creative thinking and innovation to ongoing projects. I liked the challenge. It made me feel alive. It made me feel like I was making the process, the product design, the service, the department, the company, my community, the world a little better, because I cared enough to want to make a difference.

It does not matter if your job is making widgets. Be the best widget maker you can be and be proud of your accomplishment. If your work is not receptive to your creative thinking and innovative ideas, find another place to work or apply your ideas to another facet of your life.

In this book I take the engineer and technologist through the innovation cycle by first of all preparing the mind for the process in Chapter 1 "Getting the Mind Right." I then review the history of visionaries and innovators (Chapter 2), as well as discuss an overview of idea engineering (Chapter 3). Understanding organizational roadblocks that often prevent innovative thinking and behavior is introduced in Chapter 4. Successful business models of innovation, along with options for organizing idea teams are discussed in Chapter 5. How to identify the problem to be solved (Chapter 6) is followed by a chapter on creative thinking (Chapter 7). Both chapters discuss specific techniques that have been used successfully. Next on the innovation cycle is the selection of the best idea (Chapter 8). Then the hard work begins when designing and testing and idea (Chapter 9). Societal constructs that influence the success of an innovation when it is introduced in the marketplace are discussed in Chapter 10. My final thoughts are found in Chapter 11.

Each of us has one life on this earth, and we are born to seek happiness. In order to find happiness, our lives must have meaning. Either skate through life in the state of status quo or improve the quality of your life and grow as a human being by not being afraid to learn to think differently and apply that knowledge.

The same thing applies to your career. If you are going to spend a minimum of eight hours a day working at your job, enjoy it, regardless of what you do. Hard work combined with passion for what you do generally leads to success. Be the best you can be. Make the company value your "can do" positive attitude. Take pride in your contribution. Make a difference.

Engineers and technologists often operate from a binary coded worldview of "ones and zeros." My mission in this book is to interject the colorful world of creative thinking and innovation to help engineers and technologists learn to think and work differently. This "idea engineering" becomes the driving force, transforming engineers and technologists into thinking like designers, inventers, and entrepreneurs.

So now start peeling away the layers, and get down to exposing the heart of the designer, inventor, and entrepreneur.

# Love & Gratitude (Acknowledgments)

*At times our own light goes out and is rekindled by a spark from another person. Each of us has cause to think with deep gratitude of those who have lighted the flame within us.*

—Albert Schweitzer

Love and gratitude foremost to the people who inspired me to write this book "Idea Engineering: Creative Thinking and Innovation." My special group of undergraduate and graduate students who were part of my IDEA Laboratory at Arizona State University and Purdue University were my first inspirations. I call them my IDEA Gurus. Thank you for making my academic life such a joy. I know I may have been hard on many of you, but it was because I knew what you were capable of doing and being, and I did not want you to settle on yourself. I believed in you and we had fun in the process.

My other inspirations are my grandchildren (I now have a baker's dozen), and their parents (Tonya and Steve Behnke, Rocky and Elizabeth Harris). These children are the ones who will challenge the world in the next generation. I want them to know that you can find good in this world by being part of the solution, not the problem, and that the world is full of possibilities, if you keep your mind open. I want them to know that the same thing can happen to a number of people, but it is your individual perspective that makes the event a learning moment or not.

Love and gratitude to my editor Bill Peterson, who believed in me and was the opposite of a micromanager. Bill let me run with the book and trusted me. He advised me well.

Love and gratitude to my book reviewers (Tom Raine, Rocky Harris, Dr. Jon M. Duff, Dr. Nancy Study, and Carl Harris), who knew I welcomed their feedback and I had a thick enough skin to deal with it. Love and gratitude goes to Tom Raine, business attorney at law, previously from Squire, Sanders & Dempsey, and Raine Law Firm PLLC, for being such a thorough reviewer. Tom is presently an Assistant Professor of Legal Practice at the Phoenix School of Law, and is known for his editing skills and for being a creative problem solver. Tom is a Renaissance man. He is not only a lawyer and a professor, but he is also a musician, a martial artist, and a wonderful father to his sons. I was thrilled when I heard he graduated in the top of his class at the University of Arizona law school, because I have known Tom since he was in kindergarten.

Rocky Harris (a.k.a. Brett R. Harris), Senior Associate Athletic Director, Arizona State University, happens to be my son. I actually wanted Rocky to be the co-author of this book, but he was too busy doing the things he loves to do. He is the marketing guru, a wordsmith, and the one who read the book from the viewpoint of a creative thinker. With a strong communications, English, business, and creative thinking background, Rocky is the one who read through my book for content application. He helped reorganize the content. When he left the Houston Dynamo professional soccer team as Executive Vice President to move to Arizona to be with family, Rocky gave a speech to his staff, which included his 10 Golden Rules.[1] This inspired me as I wrote my book. Even with his busy schedule, I appreciate him taking time to review my book. Dinner at my house with Liz, Hope, and Grace, when the book is published.

Thank you Dr. Jon M. Duff, Professor Emeritus, from Purdue University and Arizona State University for being one of my reviewers. Jon was one of my mentors when I went through the PhD program at the University of Arizona. He was always there by my side, not only in my graduate studies, but on a professional level in my academic career. I could count on him. I looked up to him to know the level of excellence I should be reaching for as a professor at a research university. I am forever thankful that our paths crossed. Somehow I feel like we were brother and sister in a former life. LOL. We joked about that a lot.

Thank you Dr. Nancy Study, Penn State Erie, The Behrend College School of Engineering, for dotting the i's and crossing the t's. Knowing Nancy is like having a really intelligent, caring, little sister hanging around. We met at the ASEE (American Society for Engineering Education) conferences throughout the years. A little side note here: Nancy would always rent a really fancy car, take vacation time, and drive around the city for a couple days after the conference. She likes to see the world in style. I was lucky enough to hitch a ride with her a few times in my life, as a passenger in a Jaguar in Salt Lake City, a Mustang in Portland, and a third trip that we both cannot remember.

And of course I cannot forget to send love and gratitude to my husband, Carl D. Harris. He has always been there for me in my life (as I have been in his) during the good times and the bad. Before anything goes to press, he reads it for consistency and transitional phrases, and gives the final thumbs up. I am very blessed to have a life partner who can handle me (most of the times). Thank you, Carl, from the bottom of my heart. Corny as this may seem to many, in my life I have come to realize that what really matters is to love and be loved. I have a grateful heart for each person who reads this book.

# CHAPTER 1

# Getting Your Mind Right

*Your time is limited, so don't waste it living someone else's life. Don't be trapped by dogma—which is living with the results of other people's thinking. Don't let the noise of others' opinions drown out your own inner voice. And most important, have the courage to follow your heart and intuition.*

—Steve Jobs

The foundation of learning to think like a designer, inventor, or entrepreneur is to get your mind right. Before we can begin the actual process of creative thinking and innovation, we must find common ground by going over some life lessons that have shaped my life. These are my career and life lessons I am passing on to you. Some of it will be new. Some of it will make sense. Some of it will sound very basic. Some of it will make you a bit squeamish, especially if you are not open to new ways of thinking. I have found these lessons to be important to a healthy career and life. These are not in order of importance.

## Lesson #1: Live a Life That Has Meaning

When I was a professor at Purdue University, I had the distinct honor of working with Dr. Stephen Cooper, who spent several years at the National Science Foundation, and prior to that at Carnegie Mellon University. We were in his office chatting about how life has a way of throwing curveballs, when the name Randy Pausch (October 23, 1960–July 25, 2008) came up in the conversation. Dr. Pausch was a computer science professor at Carnegie Mellon University and was diagnosed with pancreatic cancer in the prime of his career and life. Dr. Cooper was his friend and often drove him to his doctor appointments. Why in the world am I discussing cancer in a book about creative thinking and innovation? It is precisely because as hard as you try, you cannot separate your life from your career. The Last Lecture Series at Carnegie Mellon University is a forum for professors to give their final lecture, given the question: What wisdom would you give to the world if you knew it was your last chance? Dr. Pausch gave his last lecture to a crowd of 400 and published a *New York Times* best-selling book *The Last Lecture: Really Achieving Your Childhood Dreams*, which I received as a gift from my son.[1] This book you are holding in your hand (or your tablet) titled *Idea Engineering: Creative Thinking and Innovation* is my last lecture. It is the wisdom I am giving the world, if it was my last chance.

Just as Dr. Pausch talked about achieving your childhood dreams, I am focused on living a fully engaged life of possibilities sprinkled with common sense. The number one lesson in getting your mind right is to live a life that has meaning.

## Lesson #2: Common Sense and Intelligence

Common sense is not related to one's intelligence. My Aunt Suzy was incredibly intelligent. She was one of the only females accepted into the botany PhD program at Washington State University during the mid-1940s. This was at a time when women were not completing advanced degrees—especially if you were Japanese-American.

The only problem was that Aunt Suzy did not have much common sense—at least according to my mother. I was in the third grade when I realized that a smart person could do dumb things. Aunt Suzy took me to the town barber and had my long hair cut into a pixie without my mother's permission. Eureka! Common sense was different from intelligence.

How then does someone with intellect obtain the inner knowledge and judgment to do what makes sense? Does everyone have to learn the hard way through the school of hard knocks? Or is there a different approach to processing information?

Later in life, I learned that there is a scientific explanation for the difference between common sense and intelligence. A psychologist named Keith Stanovich, author of *What Intelligence Tests Miss*, catalogued 40 years of data on laboratory tasks. He narrowed them down to three functions of the cognitive mind: (1) the autonomous mind, (2) the algorithmic mind, and (3) the reflective mind. The autonomous mind is based on simple associations and repetitions of what one has done in the past. It is fast and effortless. If a person always purchases Brand X computers for the company, that is the autonomous mind at work.

Let's say for some reason Brand X is no longer available. The algorithmic mind—that step-by-step problem-solving mind—takes over and a person makes comparisons and combines the new options of computers in different ways. Standard IQ tests measure the dual cognitive processes of the autonomous and the algorithmic mind.

Common sense, however, is the reflective mind. It refers to rational thinking and is not used as frequently as it should be. Common sense can be taught to some extent by practicing ways of avoiding errors that the reflective mind makes. Simplify this way of thinking by saying that there is a shortage of funding in this year's budget to purchase new computers, so no new computers are ordered. That is common sense at work. Creative thinking starts working when ways to better optimize your existing budget are now becoming possibilities, but that is another chapter.

### Lesson #3: Stinkin' Thinkin'

Stinkin' thinkin' is a pattern of personal behavior that centers around negative attitudes. There is an Asian symbol for crisis that many believe is the same symbol for danger plus opportunity, although some dispute the old Asian cliché.

危机

So when faced with a problem look at it as being confronted with an opportunity. Holding up the white flag before beginning is a self-fulfilling prophecy. The world is full of skeptics, and they have their role as devil's advocate in problem solving. But without being open to all the possibilities, no imaginative innovation will occur.

My father told me that the only thing people have control of in their life is their attitude. Motivational speaker Zig Ziglar said the following about attitude: "Your attitude, not your aptitude will determine your altitude." I am going to have to tweak his quote and say, "Your attitude *and* your aptitude will determine your altitude." Networking connects people to opportunities, and it is who you know that gets you the job. In order to hold on to the job, however, it is what you know that counts. Then it is your attitude added to your abilities that determines how high you climb in this world.

Success comes with joyful participation and finding some kind of silver lining. In contrast, self-absorbed survival thinking is an anxious, fearful, competitive mindset of looking out for number one with the overlying cloud that life is dangerous. This is living with a dog-eat-dog mentality. This occurs when people are pitted against each other in an unhealthy manner. Without managing the fear that comes with this limiting thinking, creative thinking and the passion that comes with joyful participation in a project is nonexistent.

The ways in which we assess, measure, and compare ourselves to others is our own fabrication. We need to get rid of this conceived framework that limits us. All anyone can ask of you is to do your best with the resources at your disposal.

Finding solutions to problems requires one to think differently. Creative thinking also leads to finding solutions for problems in one's daily life—both at work and at home. When the dean approved my IDEA Laboratory at Purdue University, I was allocated one very empty room with no funding to buy furniture. Not that I was not grateful, but I could not run a research and development laboratory in an empty room. I discussed this issue with the facilities manager and he told me that when other departments purchased new furniture, all the old furniture was sent to his area for redistribution or storage. I managed to get a conference table, chairs, tables, filing cabinets, and even computers from this process. My department chair saw my efforts and allocated a couple of high-end computers for the laboratory. Then grant monies allowed for the additional computers, software, video equipment, etc. There are always opportunities, if one just looks hard enough.

Here is an example of how a limiting fearful attitude can create pain. During World War II, my dad was in the Army's 442nd Battalion, a Japanese-American battalion made up of mainly Hawaiians of Japanese descent who were fighting for the United States. Right before they were deployed overseas to fight in Germany, they were stationed in the Carolinas. One evening several members of the 442nd Battalion decided that they wanted to eat in a Southern café. These Army men were born in the United States and many had never been to Japan. They were dressed in their Army uniforms and were very respectful. They sat down at a table and were immediately asked to leave. They were not served because they were of Japanese descent. I asked my dad why he was not bitter about the experience. He said that we cannot live full and happy lives if we are bitter. And besides, he said, "It is not their fault. Some people are just plain ignorant, and they learned that from how they were raised."

## Lesson #4: Respect

A common thread that holds people together in a successful business and in personal relationships is respect. Even if you do not like a person or would never want to invite him or her over for dinner, show respect for his or her position.

I often use this analogy when talking to young people. If you are on a football team and you want your team to win, you will throw the ball to the wide receiver, even if he has been trying to move in on your girlfriend. You may not like him, but you respect his position for the good of the team.

You cannot respect others until you respect yourself. The success of your life professionally and personally depends on your sense of self-worth, and living with dignity. How do you behave when others are not watching?

Having self-respect means that you are self-confident. It is believing in your abilities. It is empowering. However, having self-confidence is not equated with having arrogance.

"Arrogance is one of the deadliest of all human failings and can destroy a business," according to Harvey MacKay, a nationally syndicated columnist and businessman.[2] Arrogance is overbearing, presumptuous, and assumes superiority. Arrogant people do not respect others.

## Lesson #5: You Are Not Superman (or Superwoman)

You are not Superman (or Superwoman), because the character is fictional and did not need a team to succeed. "You are only a human being" or "human bean" as my daughter Tonya would say when she was two. Stop trying to do everything yourself! Human beings have limitations. We need support from each other at work and play to reach our goals and live a fulfilled life.

Control freaks micromanage and do not delegate to others. If a task can be effectively completed with the technology available, control freaks will use people instead of technology, and ultimately cost the company more money. They are difficult managers, as well as co-workers, and cause resentment within the ranks of the organization. Do not be one of those. If you are, work on changing it!

## Lesson #6: Financial Literacy

A basic understanding of finance and how money works is not just a wonderful tool in making better decisions personally and professionally, but it is an indispensable one. This is a world with complex financial decisions. Without this basic knowledge, poor financial decisions that can have harmful effects on your life and your business will undoubtedly be made. Sadly enough, only half the states in the United States have required financial literacy in the kindergarten through twelfth grade school system. So if there is an opportunity to take a finance course, it will help immensely, both professionally and personally.

## Lesson #7: Failure, the Blame Game, and Perfection

A great amount of knowledge in life can be learned when things go wrong. The real success or failure in a situation is how it is handled. If you learned from falling down, and picked yourself back up, then you are successful no matter what anyone else thinks or says. So take a deep breath. There is nothing wrong with failure and making mistakes, in spite of what others believe. It is a sign of action and risk-taking. It opens up opportunities.

One of the most significant obstacles to unleashing the creative juices to solve problems is the fear of failure. Future success is possible when lessons are learned from failure, so do not fear it. Being surrounded by an organizational culture of fear is not healthy. Taking risks, and criticizing how things are done, then becomes a high-risk activity for job stability. This does not help the creative juices, people!

DO NOT play mistakes over and over in your head. It saps your energy. Understand that problems in life and at work do not define you. You do have choices. You can either remove yourself, accept it, or change it. One, two, three. Which choice will you make?

DO NOT play the blame game. This is when a person refuses to accept sole responsibility for an event that occurred and instead points their finger at others. Turn this toxic energy into energy used to find a solution.

DO NOT demand perfection and standards that are unrealistic to yourself or others. You will always be disappointed. Perfection does not equal competency. People who are hired to do a job should be expected to be competent. Competent people possess skills, knowledge, and behaviors that allow them to do their job with a degree of success. A business cannot be successful without employees who perform competently.

DO NOT focus on all the things that are wrong. You will lose sight of the big picture and you will not be effective. You will give mistakes too much power and they will evolve into a life of their own.

Instead, learn from mistakes. What will you do differently, if you are faced with the situation again? Focus on what is, not what should be. Come to the table with the spirit of openness. One of my sayings is "That's life in the big city." Or as some say, "It is what it is."

People are human and even in the best of situations mistakes can be made. Do not fight them; embrace them. When a mistake occurs in a project, it does not matter who made the mistake, just as long as we figure out a way for it not to be repeated; that is, unless the mistake turns out to be the road to a better idea. That entails some creative thinking, and that cannot take place when the environment is surrounded by fear of making a mistake.

When managers have a reign of tyranny, it is not a fun place to work. Working in a healthy environment where a person doesn't feel threatened is empowering, because one can admit mistakes without feeling like it will be used against you. Creating an environment that does not penalize mistakes allows a person to take risks and reach new heights.

## Lesson #8: Chill Out

Laughter and a great sense of humor is an open pathway to creative thinking. They both have a vital place in a creative environment. Demanding people, who feel entitled, take themselves much too seriously and need to take a deep breath. Stand back for a moment and assess your own behavior. Remember that creative thinking and innovation are difficult to cultivate in an environment full of self-pride, childish demands, and entitlement.

Rejection is a part of work life. It is important not to take it personally and to learn from rejection. I love the saying, "You don't always get a trophy."

## Lesson #9: Live in the Here and Now

Do not wake up one day and realize life has passed you by through living in the illusions of the past or the future. Living in the here and now means not worrying about what has not happened yet, and not dwelling on what has happened already. You are living in acceptance of life as it is. You are practicing mindfulness, which is a state of active and open awareness on what is happening now without judgment.

If elements of your personal or professional life need to be changed, begin in the present. Ask yourself "What do we do with our current situation?" If we are clogging up our mind with the illusions of the past or the future, there is no room to listen to what is happening today.

## Lesson #10: Perceptions of Reality

In today's global economy, understanding different cultures is a requirement. For example, being humble (*kenson* in Japanese) is a trait I learned from my dad. "One should not brag," he would say. Humility is a Japanese trait and a Christian trait.[3] It is the opposite of being proud. If you are complimented in Japan (which is a type of formality), respect must be shown for the person by deflecting the compliment. This will also help you save face. If you do not answer by indulging in self-deprecation, you will be thought of as rude. This trait helped me when I had to do business with Japanese businessmen. Humility goes hand-in-hand with a strong, but quiet work ethic. This work ethic is motivated by working hard so that your children can be educated.

From my mom, who was an immigrant from Germany, I learned that having fun was an important part of living. I believe she was born with happy genes. Her smile is what I remember the most. Laughter and resilience may have been the healthy coping mechanism for her growing up in an orphanage in Nazi Germany. Balancing both cultures in my life was not always easy, because having fun was sometimes perceived as being lazy from the serious Japanese side of the family.

I also learned from my mom that it was important to be proud of yourself—your actions and choices, and your sense of belonging; however, pride should be a virtue, not a vice.

From the American influence, I learned to stand up for what I believe in, be an individual, and take the initiative. Perhaps, today's leaders in America could develop greater trust and inspire others by the addition of a little humility and a little fun in their day.

Along with the fun, please do not be afraid of hard work. In fact, a good work ethic is what separates successful employees from ordinary employees. A person with great ideas, who does not follow through is not of much value to an innovative company.

Accept the reality of your situation and the culture of your work environment. If possible, try to find a company with a work culture that matches your personal style. In many instances in the private sector, if you are humble and never toot your own horn, you will be overlooked. Do it too much, and you will be perceived as arrogant. If you never smile, you will be labeled a curmudgeon—a disagreeable person. Life is about balance and harmony.

### *Living a Life of Possibilities*

Every action begins with your thoughts and these thoughts construct your framework of whether or not you choose to live a life of possibilities. If there are three people involved in the same situation, there are often three different perceptions of the reality of what happened. This is often called "postmodernism," but since this is an applied book, not a book of theory, that concept can be visited on your own.[4] Take a moment to think about perceptions of reality. It has amazing and powerful meaning. It is as though each person creates his or her own personal framework of assumptions. Our perceptions of the world, combined with our personality and attitude, gives us two options. We either have the tools to move forward in abundance with no fear, or stay comfortable and stagnant with the safety of the status quo. If we open up the framework of our thinking, we open up a paradigm shift of more possibilities than we ever imagined.

For example, a company sent three representatives to visit a third-world country to determine whether it would be a good market for their technological products. Representative #1 sees the low income per capita, the debt still owed by the nation, and that the over $500 billion U.S. direct aid has had very little long-term positive effect. He reports back that this country would not be a good investment.

Representative #2 observes the lack of old computer systems as a positive option to move forward without having to adapt old systems and revising data. She determines that computing power can help the people solve some of their agricultural problems.

Representative #3 sees a shift from agriculture to technology and wants to set up mobile phone technology labs. He envisions a verbal World Wide Web for illiterate people with inexpensive phones. He sees the congestion from the traffic and envisions security cameras connected to computers to determine where to build more roads.

Here we have the same situation, but three different worldviews—three different sets of possibilities.

> Art [I will substitute the words "creative thinking"], after all, is about rearranging us, creating surprising juxtapositions, emotional openings, startling presences, flight paths to the eternal.
> —Rosamund Stone Zander and Benjamin Zander

## Lesson #11: Relationship Model

Relationships are connections with people in your life and at your work. Building long-term relationships is hard work and involves compromise. If you are a cerebral person or perhaps a task-oriented person, pay closer attention to the fact that relationships are a vital part of life and work. Relationships change and time needs to be set aside to check in often with people and make notes of changing goals and expectations. Solving conflicts in a relationship can actually make it stronger. Misdirected expectations, unrealistic demands, or unresolved behaviors are usually the cause of the conflicts. Interpersonal problems cannot be resolved unless there is an air of transparency, an openness to accept another viewpoint (even if it is not fully understood), and tons of communication.

In my life, I have worked very hard at my marriage. I have respected and loved my husband for over four decades. It has not been easy at times, but it most definitely has been worth it. He and I approach life very differently, but we both end up in the same place. He is home to me.

I wanted my children to have a head start in the relationship model, so before they could be excused from the dinner table, I asked them, "What did someone do for you today that made you feel valued, or what did you do for someone?" This is a question that we need to ask ourselves more often at home and at work.

In my career and my goal for efficiency, I have often had to work really hard at making sure that I communicated more than just the facts. I learned about the Japanese management style of managing by walking around.[5] This management style was popularized by becoming "The HP Way," the open style of management pioneered by Bill Hewlett and Dave Packard, the two founders of the Hewlett-Packard computer company.

This is how they get to know their employees on a more personal level and the employees feel valued and gain trust. In my case, being half-Japanese, this concept should have rubbed off on me more easily, but often I was in a hurry to get things done. Take the time to connect with your employees, your fellow workers, your clients, and the building custodian. They are vital to your business.

In an organization, we determine the relationship model we want with the people in our professional circle. We determine what we will contribute to the relationship. We determine how we will make a difference. It all begins with trust. We cannot have trust until we have a relationship.

The relationship model is based on wanting a long-term mutually beneficial relationship with others. This trust is developed when each person feels that there is no problem that could arise that cannot be worked out. Barriers are removed and creative thinking and innovation follow.

Relationship frameworks based on short-term results cause long-term problems. When a relationship is open and the outcome is acceptance either way, we require nothing of each other. We are not manipulating. We are having a healthy relationship.

## Lesson #12: Buy-in and Mutual Benefits

When your mind, body, and heart believe in the idea, there is buy-in. This passion and participation are only realized when a contribution to the idea has been made and it was valued and heard. Becoming motivated and engaged leads to wanting others to be a part of this project. Remember that you did not get this way by being manipulated into participating. Buy-in to the idea should be mutually beneficial. If people do not see a benefit, then all you are doing is persuading them to do what you want. Show them their benefits.

The wonderful thing about buy-in and mutual benefits is that there is no "us" versus "them." There is no one to blame. It is a common vision toward a common goal, and everyone is committed. It is a magnet for inspiration. You will soon be hearing "What is the best decision for US?" "Which direction are WE going in next?" We are now connected and a team. That is a healthy place to work.

Some environments are not as healthy as others. I have worked in healthy and toxic environments. One of the most important things to learn about people in order to survive in any environment is to find out what motivates them. Is it philanthropy? Money? Recognition? Power? Opportunity to be a part of something extraordinary? Being valued?

I was a manager overseeing a creative and technical department. I had recently hired a database administrator (I will call her Ella). She was a transfer from the sales and marketing department where she was the administrative assistant under a charming, but overbearing boss. The reason I was able to hire Ella was because she had made an effort to go back to college and take courses so that she could qualify for the database position. Not only did I respect her for her efforts, but I also believed she was the best qualified candidate for the position.

After a couple weeks on the job, Ella's former boss stopped by her desk and demanded that she serve food at a marketing event to be held the next evening, just as he would have done months ago when she was still his administrative assistant. When he left, I saw that she was crying and understood that she was rather intimidated by him. She had worked hard to earn another position in the company and had deadlines that had to be met in our department, but she did not know how to say "No" to him.

It did not bother the sales manager or the general manager that she felt devalued because of the request and felt powerless. I even offered the use of the administrative assistant on my staff, who was quite excited about participating in the event. They said "No." They were not sensitive to Ella's feelings. They were not motivated by my department's deadlines, that is, until they saw a direct correlation to the impact on the sales and marketing department.

I knew that the only way I could plead Ella's case was if I targeted whatever motivated the sales manager and the general manager. When I realized that they were motivated by the budget, I explained that the three hours she would be gone from her database responsibilities was going to cost the company X amount of money. We had a major project due the next day and if she did not do her part, I would have to send the computer graphic artists and writers home for those three hours, and then bring them back at time-and-a-half. It was going to cost the organization thousands of dollars. I requested the money be transferred to my department budget so that the project could be completed in time for the client. They revisited my offer to use my administrative assistant and suggested that as the solution. I agreed.

Sometimes we have to play stupid corporate games... and then we have to rewire our heads to get our minds right.

## Rewiring Your Head

This is the part of the book that may make some black and white thinkers[6] feel very uncomfortable and claim that this section is a bit peculiar. That is the purpose. This is when you open your mind and commit to thinking not only out of the box, but also under it, through it, and hovering above it ... or perhaps you should just redraw the box.

Rewiring the way you think is not an easy undertaking. One cannot think creatively and be innovative, if one does not de-stress and begin with a firm foundation. You are essentially reprogramming the thoughts you allow into your brain so that your actions reflect that inspiration. Be persistent and honest to yourself so that your positive intentions are not merely masking your negativity. It is important to truly free yourself of the negative records that are played repeatedly in your head, because negativity can keep you from living the life you deserve and having the career you desire. Be sure your thoughts do not come from a place of fear.

### Calming Your Mind

Meditation takes many forms and calms your mind. As a part of my daily routine, I meditate about 10 minutes once or twice a day. Build up to 30 minutes a session, if you choose.

Meditation shifts your brain activity from the right frontal cortex (which is stress-prone) to the left frontal cortex (which is calm). The right frontal cortex is also the part of the brain where neuroscientists say fear is processed. By making this shift in brain activity, the negative effects of stress, anxiety, and mild depression are decreased. Your blood pressure can even be lowered through this activity.[7]

I enjoy doing Tai Chi with my friend Sue Garcia, along with my daily ritual of calming myself in the morning and evening. It is a form of meditation and keeps me centered.

Some people like my daughter-in-law Elizabeth take up yoga. If this is something you would like to try, I would recommend a newly published book *Yoga Wisdom at Work: Finding Sanity Off the Mat and On the Job* written by Maren and Jamie Showkeir.[8] Maren and I worked together at *The Arizona Republic* years ago.

### Visualization

Research suggests that the practice of visualization, which is guided imagery, increases creative thinking and decreases stress. It is a type of meditation.[9]

Visualization can be done in silence, while listening to instrumental music or repeated sounds, such as ocean waves. Close your eyes. Imagine you are in one of the places that is the most relaxing for you. Smell the air. Observe the beauty. Listen to the sounds. Feel the breeze against your skin. Taste the salt in the air. Make all your other thoughts STOP.

Begin with 10 minutes of visualization every morning or every evening and do this for one month to start. Move up to 20 minutes (or 10 minutes in the morning and 10 minutes before you go to sleep). Clear your mind. Sit on the floor or stand in a quiet place, and relax your body, shoulders, and arms. And breathe in with your nose into your belly, and out with your mouth.

### *Vision Boards*

Another option to relieve stress is to make a list of the positive things you want to attract in your life. Sit or stand in a quiet place, and relax your body, shoulders, and arms. Make a mental vision board of all those positive things. As you breath air in with your nose filling your belly and out slowly with your mouth, use your senses to make your visions believable.

### *Music Meditation*

Music meditation is simply taking the time to sit or lay down quietly with or without closing your eyes and listen to soothing music.

### *Repeated Words of Affirmation*

Thoughts and beliefs become words. Because words become actions, repeating a positive word or phrase is another option, if you have difficulty quieting your mind. I do this when my mind starts wandering and it works very well. Repetition of the positive phrase becomes a mantra for transformation. "May I be free from …" "I choose to …" "I will…" "May I accept…" "May I receive…" "I empower myself to attract …" Add a specific timeline. Repeat the same phrase for at least 10 minutes.

The subconscious mind responds to repetition. This is what quantum physics, which is the study of energy and matter, refers to as the

Law of Attraction.[10] This means that you attract into your life whatever is manifested in your thoughts. And if you think this is a bunch of hooey, I am telling you that it will not hurt you, if you at least give it a try.

## Food for Thought

So if your mind is right, if the meditation has helped you embrace thoughts of gratitude and love, and if you are now mindful, we can move on to learning about defining idea engineering.

# CHAPTER 2

# Visionaries and Innovators

*The visionary starts with a clean sheet of paper, and re-imagines the world.*

—Malcolm Gladwell

This chapter discusses the differences between visionaries and innovators and reviews the lives and contributions of several of yesterday's and today's visionaries and innovators. The names Marie Curie, Thomas Edison, Albert Einstein, and Leonardo da Vinci are some of the creative thinkers of the past who come to mind. They have inspired modern-day thinkers, including Patricia Bath, Mark Zuckerberg, Jeff Bezos, Bill Gates, and Steve Jobs. This chapter is important because respecting and understanding the journey of visionaries and innovators of the past and present give us a foundation to build upon.

This chapter is also important because engineers and technologists get so wrapped up in the production, building, and implementation of an idea that they do not spend enough time being a visionary. Settling on the first workable idea that comes to mind results in mundane products, processes, and services.

## Visionaries Versus Innovatiors

Being a visionary and being an innovator are not the same, but the terms are inter-related. There is no innovation without vision.

Visionaries are creative thinkers, who generate novel ideas and improvements to solve a problem. A visionary uses creative thinking to **imagine** the product, process, or service. A visionary is often referred to as an "idea person."

Innovators are trailblazers. Yes, they are visionaries, but in addition, they **apply** their novel ideas and improvements to produce, build, or implement a concrete product, process, or service.

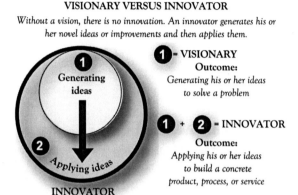

**VISIONARY VERSUS INNOVATOR**
*Without a vision, there is no innovation. An innovator generates his or her novel ideas or improvements and then applies them.*

## Visionaries

A visionary creates ideas and actually imagines change. A visionary is a dreamer of possibilities—people with revolutionary ideas. A visionary also has the wisdom of having keen foresight as to what lies ahead. A visionary with a great imagination and big ideas that sometimes appear crazy or impractical to others has the potential to change the way the world works. Great leaders are visionary thinkers.

True visionaries inspire. Visionaries are creative thinkers, who come in many packages. This means that society should not prejudge the person who comes up with the next crazy idea. It could be the fearless 13-year-old dreamer next door or the 90-year-old great grandmother.

## Innovators

An innovator follows through on ideas and actually produces, builds, or implements change. An innovator takes his or her new ideas or alternative ways of doing things and applies ingenuity to develop tangible solutions to identified problems. An innovator is also a visionary who uses his or her thought processes to think differently. This leads to pioneering novel ideas or making improvements on established products, processes, or services.

Trying out entrepreneurial ideas requires the ability to take risks and it requires follow-through. Innovators are not afraid of failing—in fact, they welcome it. Fear is risk averse. The benefit of failure is that it shows what will not work and takes you a step closer to the solution. So embrace it! San Francisco's Silicon Valley is ripe with failures and therefore with big ideas that have been applied. Other regions in the world tend to be more risk averse and therefore not as entrepreneurial.

## Rube Goldberg

No book about creative thinking and innovation should be written for engineers and technologists without the mention of Reuben Garrett Lucius "Rube" Goldberg (1883–1970). Goldberg was an engineer-turned-syndicated cartoonist. The cartoon of Professor Lucifer Gorgonzola Butts was Goldberg's legacy for engineers and technologists. The Professor Butts cartoon consisted of labeled schematics of inventions, which accomplished a simple task through complex means. Today, the term "Rube Goldberg" is defined by *Webster's New World Dictionary* as "a comically involved, complicated invention, laboriously contrived to perform a simple operation."[1] Rube Goldberg's official website and gallery of images is available online at http://ww.rubegoldberg.com.

Goldberg's art and contraptions have inspired the imagination of many engineers and technologists. People embrace the inventiveness of Rube Goldberg machines by participating in contests open to all ages throughout the United States and internationally.

The first Purdue University Rube Goldberg machine contest took place in 1949 stemming from a rivalry between two engineering fraternities. The contest was revived again in 1983. In 2012, Purdue University beat its own Guinness world record for the largest Rube Goldberg machine with 300 steps that accomplished the task of blowing up a balloon and popping it.[2] Prior to that record, the Purdue University 244-step time machine claimed the Guinness world record in 2011.[3] The winners have made appearances on television venues such as *The Tonight Show*, the David Letterman show, CNN, *Jimmy Kimmel Live*, *Newton's Apple*, and NBC's *The Today Show*.[4]

---

**TRAITS OF VISIONARIES AND INNOVATORS**

*The following lists are traits that have been found in the research. One can be an idea person without being an innovator, but without ideas there is no innovation. An innovator is a visionary who applies his or her ideas.*

*Not all of the traits are possessed by each, and many of the traits overlap.*

| A visionary … | An innovator … |
|---|---|
| Is a dreamer | Is a trailblazer |
| Is a creative thinker | Is a visionary who applies his or her ideas |
| Has an open attitude | Produces, builds, or implements change |
| Inspires people | Is not afraid to fail |
| Has keen foresight | Is a risk-taker |
| Has a vision for what consumers want | Is resilient and does not give up |
| Has imagination | Is persistent |
| Is resilient | Is committed |
| Faces life adversity successfully | Is a lifelong learner |
| Has been mentored by someone | Is sometimes philanthropic & generous |
| Has a joy of learning and understanding | Has follows-through |
| Is intelligent | Has inventive behavior |
| Has a quest for knowledge | Is not afraid of hard work |
| Is educated and/or a lifelong learner | Is strategic |
| Loves to experiment | Is patient |
| Is curious | Is often competitive |
| Is clever | Is driven |
| Is energetic | Often has an innate business sense |
| Is passionate | Is committed to his or her work |
| Has intense drive | Is courageous |
| Has a good work ethic | Collaborates with others |
| Has a diversity of interests | |
| Is sometimes unconventional | |

---

# Idea Engineers of the Past

## Marie Curie[5]

Marie Curie (1867–1934) was born Maria Salomea Sklodowska. She was a Polish chemist and physicist, and one of the most prominent female scientists of the past. Curie discovered polonium and radium and was the first person in history to be awarded two Nobel Prizes for her innovative work in radioactivity. This led to a new epoch in the treatment of diseases.

In spite of life's challenges, many visionaries like Marie Curie are resilient and focused on their goals. Because of her family's political pro-Polish beliefs, they struggled financially. Warsaw was under the rule of the Russian Empire at the time, and her father was forced out of his teaching job.

Before Curie was 11 years old, she was faced with the death of her sister, who had died of typhus, and her mother, who had died of tuberculosis. Four years later (at the age of 15), Curie managed to graduate from high school with the highest honors.

Shortly after that, Curie suffered from what modern-day physicians would diagnose as depression. Her father sent her to live with the family in a nonstressful environment in the countryside for a year.

At that time, it was illegal for Polish youths to attend university-level courses at the University of Warsaw, but that did not stop Curie. She was driven by her joy of learning and understanding. She joined her older sister Bronya and other young Poles in secret underground higher-education studies—a "floating university." They met and studied at night and continued to change locations of their meetings to avoid being caught by the czar's police. Curie found access to a laboratory and this confirmed her passion for experimental research in physics and chemistry.

Curie worked as a governess to pay for her sister's medical studies in Paris. The plan was that once her sister started earning money, she would pay for Curie's tuition, a small room, and the bare minimum of food.

Even though Curie was not as academically prepared as the other students, she had a good work ethic and managed to finish her master's degrees in physics and math in three years. She was offered a scholarship by a group of industrialists. The freedom to explore the new academic world gave Curie a sense of liberty.

### *Thomas Edison*[6]

Thomas Alva Edison (1847–1931), inventor and American entrepreneur, was the youngest of seven children. He is best remembered for his inventions of the motion picture camera, and the phonograph, and the success of his electric light bulb for home use.

I have researched Edison, and there are several characteristics that define him as a visionary and an innovator. Edison was faced with many life challenges and managed to view them as inconveniences, not obstacles that could not be faced successfully. One of his challenges was his health. He had poor health as a child. He was also delayed in speaking and had difficulty with words. Edison did not start talking until he was four years old, and when he did, his voice was noticeably high pitched. In addition, when he was about 12 years old, Edison lost most of his hearing. This was due to childhood scarlet fever and untreated middle-ear infections.

Edison was also very curious and energetic. Like many children, Edison was always asking the question "Why?," but this curiosity continued into adulthood. This characteristic is a strong indicator of a creative mind.

A high energy level is also a characteristic of an innovator; however, Edison's energy level was a little extreme. He was passionate when interested, but his attention was easily diverted. Because of this, he got into mischief. He spoke at inappropriate times and did not listen to the teacher. Many experts have since speculated that had Edison been a child in modern times, he would have been diagnosed with attention deficit hyperactivity disorder (ADHD).[7]

Edison also lacked patience and was rather clumsy, resulting in him not being very athletic. All of these challenges as a child resulted in Edison being incredibly shy.

The one strong foundation that grounded Edison was his mother. She was committed to him and his success in life. Because of all his childhood health issues, Edison was a poor student. His mother pulled him out of school and taught him at home. In spite of all of his life challenges, Edison became one of the most revered inventors.[8]

### Albert Einstein[9]

Albert Einstein (1879–1955), a professor, scientist, and inventor, was born in Ulm, Germany. Today, Einstein is considered the most influential physicist of the 20th century. He is famous for his theory of relativity.

Einstein was awarded the Nobel Prize in Physics for his explanation of the photoelectric effect. His work laid the foundation for the inventions of DVD players, lasers, digital cameras, remote control devices, televisions, automatic door openers, and so on.[10]

Einstein had many characteristics that are found in visionaries. He was educated and had a variety of interests. Einstein was an excellent student and wrote his first scientific paper at the age of 16 on magnetic fields and light. In 1905, he earned his doctorate degree and published the theory of relativity, but science was not the only interest he had. As a child, he took violin and piano lessons.

Even though his life began quite comfortably in an educated middle-class Jewish family, Einstein struggled with a delayed speech impediment and dyslexia. He spoke very slowly as he had to get words correctly in his head before he spoke them out loud. He did this until he was about nine years old. Einstein's dyslexia was the cause of his inability to remember the simplest things, such as how to tie his shoelaces and the days of the month; yet, he had no problem solving some of the most intricate mathematical formulas.

One of the predominate characteristics of an innovator is having persistence. Einstein was a high-school dropout for political and social reasons. He ended up applying to the Polytechnic Institute in Switzerland. He got exceptional marks in physics and mathematics, but failed much of the rest of the entrance exam. He was admitted with the requirement that he get his high-school diploma first. Einstein did so at age 17 in Switzerland.

Einstein was a war pacifist and believed in social justice. He dreaded his required military duty in Germany, and that is why he ended up dropping out of high school and becoming a draft dodger with no skills for employment. Ironically, Einstein wrote a letter to President Franklin Delano Roosevelt encouraging the United States to build an atomic bomb to counter Nazi Germany's potential research in building an atomic bomb.

When Einstein graduated from the Polytechnic Institute, he had difficulty finding a job because his professors blackballed him from every academic position to which he applied. Einstein was misunderstood. He had a reputation for cutting classes—not because he did not want to go to school, but because he preferred to study on his own.

Einstein was faced with many obstacles, but that did not stop him from succeeding. His family had gone bankrupt, he had a child out of wedlock, and Einstein finally ended up as a clerk in the Swiss patent office. Einstein's life at that point was desperate. The job came so easy to him that he was left with lots of time to work on his inventions.

### Leonardo da Vinci[11]

Leonardo da Vinci (1452–1519) was an Italian artist, mathematician, musician, scientist, and inventor. He was a master Renaissance painter; yet, he left the world with only a handful of completed paintings, such as the *Mona Lisa*, *The Last Supper*, and paintings on the Sistine Chapel.

There are several characteristics about Leonardo that made him a visionary of his time. He was a fervent observer of his surroundings and had a plethora of ideas. Leonardo documented and sketched an amazing number of his observations and ideas in his sketchbooks. He had a scientific approach to his ideas. His sketchbooks were filled with drawings of linear and aerial perspective, illustrations of human and animal anatomy, the patterns of botany, and mechanical sketches that were combined and modified to create inventions. Some of the many inventions Leonardo documented and sketched were the following: ball bearings, parachute, ornithopter (a flying machine with bat-like wings), a 33-barrelled organ (similar to a machine gun), a diving suit, an armored tank, a self-propelled cart, a planned city of the future, and an aerial screw (similar to a helicopter).

Leonardo was unconventional for his time and a bit controversial. These are not necessarily characteristics of a visionary or innovator, but what it implies is that he was not afraid to risk being different from everyone else in his society. That lack of fear is definitely a characteristic of an innovator. Leonardo was a vegetarian, animal lover, and left handed, which was an unusual combination at that time. He was against war; yet, in his sketchbooks, inventions of deadly weapons were sketched in detail.

He had one trait that was not characteristic of an innovator. Either Leonardo had a lack of follow-through or he did not live long enough to apply his ideas to inventions. Leonardo never published his ideas, and none of his inventions got built during his time, except for a robotic knight that he built himself that was used primarily for entertainment at parties for his wealthy patrons. The robotic knight was operated by a system of pulleys and gears. It was capable of walking, sitting, and moving its jaw. In 2002, NASA saw the usefulness of Leonardo's robotic knight in the aerospace industry. They used some of Leonardo's robot design to build planetary exploration robots.[12] NASA also credited Leonardo with solving the mystery of earthshine—when at sunset, a crescent moon appears on the horizon with an image of the full moon.[13]

## Modern-Day Visionaries and Innovators

I selected five modern-day visionaries: Patricia Bath, Mark Zuckerberg, Jeff Bezos, Bill Gates, and Steve Jobs. If you are wondering why four of the five modern-day visionaries are from the technology world of computing, it is because technological innovation from the 20th and 21st centuries is directly linked to the economic success of the United States, which is directly linked to engineering innovation. One of the most significant U.S. economic exports may be technological innovation. Read more about this in the book *A Century of Innovation: Twenty Engineering Achievements That Transformed Our Lives* written by George Constable, Bob Somerville, and Neil Armstrong.

### *Patricia Bath*[14]

Patricia Bath was born in 1942 in Harlem in New York City. Her father was the first black motorman for the New York subway system, and her mother was a housewife and domestic worker. Bath invented the Laserphaco Probe, which is a laser technology device that treats cataracts and can restore the sight of people who have been blind for more than three decades. In 1988, she became the first African-American female doctor to be awarded a medical patent.

In spite of the modest means of her family, Bath had a solid family support system. Both of her parents valued education. Her mother Gladys, who descended from African slaves and Cherokee Native Americans, used her salary as a domestic worker to save for her children's college education. Gladys encouraged her children to read. As a young girl, Bath was given a chemistry set by her mother to encourage experimentation and learning. Her father Rupert, who was a well-educated merchant seaman, valued the discovery of new cultures and influenced Bath to be curious. Occasionally, her father wrote a column for the local newspaper.

One of the characteristics of Bath was her intense drive and focus on her goals. She was the editor of her high school's science paper. When she was 16, she was one of the few students in the United States who was selected to go to a National Science Foundation summer program at Yeshiva University in New York. She was invited to study cancer with leading U.S. researchers. At the end of the program, Bath had developed a mathematical equation to predict the rate of growth of cancer. Her work was incorporated into a joint scientific academic paper that was presented at an international conference in Washington, D.C. She was able to graduate from high school in only two-and-a-half years.

Bath had many diverse interests. She spoke French, played the flute, and mastered English literature. Bath took art in a summer program in Yugoslavia, which focused on pediatrics research.

She also pursued her education with a vengeance. After graduating from Hunter College in New York with a Bachelor of Arts degree in 1964, she was accepted in medical school at Howard University in Washington, D.C. Her belief in the leadership role of African-Americans grew with her exposure to the professors and administrators at Howard University.

As a visionary and an innovator, Bath believed in giving back to her community. After working as an intern at Harlem Hospital and completing a fellowship at Columbia University, she recognized that there were twice as many African-American patients suffering from blindness than Caucasian patients. Her research led her to discover that the poor suffered from blindness due to lack of access to eye care. Bath was the first African-American to finish a residency in ophthalmology in 1970 from New York University. She established a new discipline called Community Ophthalmology. Bath was passionate about providing vision, cataract, and glaucoma screening to the poor and the elderly, as well as providing glasses for schoolchildren. She then became a professor at the University of California, Los Angeles (UCLA) and continued until her retirement in 1993.

VISIONARIES AND INNOVATORS    27

As a visionary, Bath made a connection between diverse disciplines. She saw the merit in the collaboration between technology and medicine and advocated using technology to provide medical services in isolated areas (telemedicine).

## Mark Zuckerberg[15]

Mark Zuckerberg (born in 1984) is the youngest of the visionaries I reviewed. He is famous for the development of Facebook, Inc., as a social networking website.

One of the characteristics of a visionary is showing signs of curiosity and inventive behavior in his or her youth. In high school, Zuckerberg invented a music player (Synapse Media Player) using artificial intelligence to gather the listener's music choices. Zuckerberg was recruited by America Online (AOL) and Microsoft, and the companies tried to purchase his music player. Zuckerberg went to Harvard instead, and uploaded the app for free.

One of the recurring themes in famous visionaries is that they have been mentored. His father taught him Atari BASIC programming in middle school, and Davis Newman, a software developer, was later hired as a private tutor.

Zuckerberg was highly energetic and intelligent and managed to be one of the five co-founders of Facebook, Inc., which was launched from his dormitory room at Harvard University. He was a risk-taker. Zuckerberg ended up dropping out of his sophomore year so that he could focus full time on Facebook.com.

## Jeff Bezos[16]

Jeff Bezos (born 1964) is Amazon.com's founder. He moved from selling books online to having his company become the world's most successful online retailers. Bezos also introduced the world to products such as Amazon Kindle (digital books) and the public cloud innovation, saving web-based businesses billions of dollars associated with business operations maintenance.[17]

Bezos appeared conventional at first by entering college, completing his coursework, graduating from Princeton University, and then working in Wall Street in the computer science field. Before his 30th birthday in 1994, however, he quit Wall Street and relocated his office to his Seattle garage to sell books online. This may have appeared to be an unconventional and a risk-taking move. But Bezos was strategic and applied his computer science and business skills to become the world's most successful online retailer. He was also persistent and patient. Amazon.com went public in 1997, but did not turn a profit until late 2001.

Another visionary trait is a diversity of interests. Bezos has invested in future innovation. His vision is more than online retail. Bezos has invested in Blue Origin, a secretive aerospace company and the building of a 10,000-year-old clock, which will be carved in one of the mountainsides of the Sierra Diablo Mountain Range in Texas.[18]

## Bill Gates[19]

Bill Gates (William Henry Gates III) was born in 1955 and is cofounder of Microsoft, valued at over $40 billion. He is the chief visionary for the American multinational software corporation located in Redmond, Washington. Gates is one of the most famous entrepreneurs of the personal computer revolution.

Microsoft dominated the personal computer operating system market with MS-DOS in the mid-1980s, followed by Microsoft Windows. Microsoft Office software is used throughout all computer platforms. The company has diversified and added Skype Technologies, Bing (Internet search engine), Xbox, mobile phones, and tablet computers to its collection.

Gates grew up in a warm and close upper-middleclass family as the son of an attorney. His mother would often take Bill with her when she did volunteer work in the community. His family encouraged competitiveness and a quest for excellence. Gates was athletic and also loved playing board games.

Gates loved to read, but when he was 11 or 12, his family became concerned with his withdrawn behavior. Even though Gates' family was a strong supporter of public education, they decided to enroll him in an exclusive private school. He bloomed there. He was known to be one of the smartest students at the school and was able to skip math. Gates even did well in his drama class.

Gates had a natural insight for computer programming. At the age of 13, he spent most of his free time on the computer with his good friend Paul Allen, who was two years older. During his first year at the private school, Gates wrote his first interactive computer program—a game of tic-tac-toe that allows the user to play against the computer. Gates and Allen were often combative in their relationship and often had their computer privileges revoked at school because of bad behavior. Gates and Allen even hacked into a computer company.

Even at a young age, Gates was driven and was a competitor. In 1970 at the age of 15, Gates and Allen earned $20,000 developing Traf-O-Data, a computer program that monitored Seattle traffic.

Gates graduated from high school in 1973, scoring 1,590 out of 1,600 on his SAT test, and was sent to Harvard University to become a lawyer like his father. Gates spent most of his time in the university computer lab instead of in class. He would sleep only a few hours, cram, and pass his tests with acceptable grades. Gates' quest for knowledge was on his terms. He did not complete his university degree, dropping out of his junior year to devote his time to Microsoft.

Another characteristic of Gates was that he had an innate business sense and knew how to stand his ground. He realized early in the game that continued distribution and use of software without paying for it was unethical and not good for business. At the age of 23, Gates was head of Microsoft, which grossed $2.5 million.

Employees of Microsoft were often confronted by Gates to push for more ideas and to think more creatively. He did not ask them to do anything that he was not willing to do. He was passionate and committed to his work. Gates was even found sleeping under his desk in the wee hours of the morning after working through the night.

Gates is persistent and courageous. Microsoft was faced with an antitrust lawsuit in 1998 by the U. S. Justice Department. Bundling Windows with Microsoft's Internet Explorer browser was viewed as anticompetitive behavior. Microsoft appealed the initial court ruling against the company and won. Gates was able to contend that the two products were the same.

In spite of his innate business shrewdness, Gates is a generous and philanthropic man. What is most inspiring is how Gates is using his wealth and his quest for knowledge to make the world a better place. He recently invested in a nuclear reactor to burn uranium in a safer, cost-efficient, and greener way. The Bill and Melinda Gates Foundation, founded in 2000, was formed to wipe out poverty and some of the world's deadliest diseases.

As a lifelong learner, Gates is inspired by a 500-year-old manuscript penned by Leonardo da Vinci that he purchased for $30.8 million. Gates' interview about the world's most valuable historic document was broadcast on television's *60 Minutes* show. Gates says that Leonardo was a great thinker well ahead of his time and had an amazing understanding of science. This quest for knowledge is what motivates a visionary.

Gates sees the value of investing in collaboration with other innovators from diverse backgrounds. Gates' project ResearchGate is an online collaboration for researchers focusing on medicine, computer science, and biology. Currently, it has 2.8 million members mainly from the United States, India, the United Kingdom, and Germany. The collaboration between researchers in Nigeria and Europe resulted in the discovery of a lethal pathogen that has the ability to cross over from plants to humans.

## *Steve Jobs*[20]

Steven Paul Jobs (1955–2011), renowned technology pioneer, became a millionaire at the age of 23. He is best known for being the chief visionary for Apple Corporation with his revolutionary personal computer and his iconic Apple brand.

Jobs began his life untraditionally. He was born in San Francisco to Abdulfattah Jandali and Joanne Carole Schieble, an unmarried couple. He was adopted by Paul and Clara Jobs. He grew up in the suburbs of Mountain View, California.

When Jobs was in high school, he worked as a summer intern at Hewlett-Packard and met Steve Wozniak, his future business partner. Jobs graduated from Homestead High School in Cupertino, California, and then enrolled in Reed College in Oregon. Jobs had diversified interests. Jobs only attended one official semester of college studying literature, poetry, and physics. He audited a number of courses, including calligraphy, which he credits with inspiring the Apple typefaces. He then left Oregon and worked briefly in 1974 as an engineer for Atari in California, teaming with Wozniak to help develop hardware required for a single-player version of *Pong*, a breakout video game. Jobs admits that Steve Wozniak was the first person he met who knew more about electronics than he did.

As a visionary and an innovator, Jobs was unconventional, a risk-taker, and focused. In 1976 after Jobs traveled through India and returned to California to live in a commune, the Jobs–Wozniak team used Jobs' family suburban garage as their business location to build personal computers, which they sold to a local electronics store in Mountain View called the Byte Shop. In 1977, Apple Corporation was founded when Jobs was 21. The Apple logo was created to represent Jobs' favorite fruit with a "byte" taken out of it. Wozniak controlled the design side and Jobs controlled the business side.

Jobs was driven, passionate, and had a vision for what consumers wanted. Jobs and Wozniak coinvented the Apple I and II computers with others. The Apple II was released in 1977. It was the first commercially successful personal computer in a plastic case with color graphics. A few years later, Apple went public with a net worth of $200 million. Two years after going public, Apple became a Fortune 500 company, and John Sculley, head of PepsiCo was recruited by Jobs to be the CEO.

Jobs had the reputation of not always doing the expected, and this caused some personality conflicts. Jobs had formed a creative team in another building of the company adorned with a pirate flag. In 1984, the team invented the Apple Macintosh computer, the first home computer with a graphical user interface that was mouse driven. It had glowing reviews but less-than-expected sales. A power struggle resulted between Sculley and Jobs and the Board of Directors. Team members complained about Jobs being demanding and rude. Sculley attempted to force Jobs out, and the board sided with Sculley. In the spring of 1985, they stripped Jobs of his Apple chairperson responsibilities and gave him an office that was referred to as "Siberia." In the summer of 1985, Jobs resigned. He was 30 years old.

Feeling betrayed and still having a chip on his shoulder, Jobs sold $70 million of his Apple stock to start a new venture NeXT Computer, a high-end computer company, which was later named NeXT Software. He was free to start all over again. He did not lose faith. He kept on doing what he loved, and money followed. A decade later (1996), Apple bought NeXT, and Steve Jobs returned to Apple as CEO.

In addition to his new venture in high-end computing, Jobs bought the computer graphics division of Lucasfilm Limited in 1986 and started Pixar Animation Studios. The original Pixar was supposed to be a high-end graphic hardware developer, but the goal was revised to be animated films. Pixar produced films such as *Toy Story* in collaboration with Disney. This led to Disney purchasing Pixar from Jobs in 2006.

After returning as CEO, Jobs led a team of innovators in product development with the iMac, iPhone, and iPad, among others. He retired in 2011 because he had pancreatic cancer. Jobs was credited with the invention or coinvention of 342 U.S. patents. The day before he died, Jobs was issued a patent for the Mac OS X Dock user interface.

To learn more about Jobs' philosophy, watch Jobs 2005 Stanford University graduation speech on YouTube.com. Through his speech, his persistence and his resilience was evident, and it shows why this extraordinary man was such a visionary.

## Food for Thought

So what makes people visionaries and innovators? Is one required to have an Ivy League university degree or be born with an extraordinary IQ? Absolutely not. Creative thinking and innovation can be learned skills. Edward de Bono says that in spite of the wild ideas that often come out of creative thinking, the process is to be taken seriously.

How do we engineer ideas, create new value, and make the world a better place? As innovators, we are accountable for enhancing our work, our play, our society, and the world with products, processes, and services that are technically, socially, aesthetically, culturally, and environmentally responsible. To have any hope for a better world, each individual must begin by taking responsibility to improve himself or herself. Inspired people, who feel a responsibility for humanity, are the future creative thinkers and innovators.

Positive, forward-looking people with open attitudes have the best potential of becoming visionaries. Great visionary leaders invest time in their teams. They are inspirational and passionate entrepreneurs, who communicate the vision, the challenges, and the goals and explain how critical each employee's role is in reaching the vision successfully. When each team member buys into the vision, and shares a common goal, he or she is more likely to accept any difficulties or changes that occur on the journey. Visionary leaders delegate responsibility and trust and empower staff by giving up control over their work. This leads to workers becoming better creative thinkers.

Successful innovators are hungry. They are driven. They are committed. They collaborate. They are respectful. They share the credit. They have an incredible work ethic. They have fun. As a professor at a research university, and as a manager and director in the private sector, I can spot this hunger in a university student and an employee. These visionaries and innovators are treasures to me. They are the hope of the world. They are our future. Those I have met in the journey of my life are permanently etched in my heart and I carry them with me at all times.

After centuries of creative thinking and innovation, it is regrettable that there are not more women and minorities in the mix. Perhaps universities and the private sector will continue to do more to attract diversity, because I believe amazing ideas are born when people with diverse backgrounds come together to think like designers, inventors, and entrepreneurs.

# CHAPTER 3

# Idea Engineering Overview

*Necessity is the mother of invention.*
—Plato

The attributes of the engineer and technologist of the future include practical ingenuity and creativity. When we talk about the "creative engineer" or the "creative technologist," others who are not engineers or technologists think those terms are oxymorons. In the new global economy, the attitudes will have to evolve. But engineers and technologists are often viewed as traditionally "black-and-white" thinkers, because they get too focused on the limitations of design. Creative thinking is often ambiguous, and ambiguity is "gray."

In this chapter, I discuss some basics about creative thinking and innovation. I then move on to discuss why we innovate and the economy and innovation. I end with the need for training and education.

The takeaway from this chapter is that innovation and problem solving cannot occur without creative thinking, because creativity is at the heart of innovation. The new reality for prosperity is indeed innovation, and that is possible through what I call "idea engineering." This engineering of ideas that transforms concepts into useful products, processes, and services creates value for stakeholders, drives economic growth, and makes our standard of living much better. So, for engineers and technologists to become more innovative, they need to learn about what makes designers, inventors, and entrepreneurs tick.

## Creative Thinking Versus Critical Thinking

Mason Cooley has been quoted as saying, "Art begins in imitation and ends in innovation."[1] Creativity is the abstract process of thinking of ideas, whereas innovation is the application of useful ideas.

What exactly is an idea? It is a specific thought or concept that arises in the mind as a result of divergent thinking[2] and also of making connections between sometimes totally unrelated ideas. An idea may be pure invention, a different way of doing things, or an improvement on an existing product, process, or service. Often, ideas are mistakenly judged by how novel they are, rather than how useful they will be to the organization or the person. By itself, an idea is not enough. It may be only a scrunched-up napkin with sketches on it.

## Creative Thinking

Creative thinking—seriously learning how to think like a designer, inventor, or entrepreneur—can lead to innovation. With the right attitude, a little practice, and a renewed sense of shedding suppressive learned behavior, everyone is capable of thinking creatively. To improve creative thinking, one needs to learn to accept change, be more flexible, be more receptive and playful with possibilities, be appreciative of the good things in life, but always strive for improvement. For some, this comes easier than for others.

Creative thinking is a right-brain activity diffusing one of the many solutions to a problem through subjective associative thinking.[3] Creative thinking is used to generate and suspend judgment on possibilities, usually through visual means, by completing the statement "Yes, and ... " An example is "Yes, that is a possibility, and by adding the fan it will cool down the battery." The creative thinker's lateral thinking[4] is divergent and is concerned with seeking as many different approaches as possible to change concepts and perceptions.[5]

## Critical Thinking

Critical and creative thinking are not the same. Critical thinking is a left-brain activity that focuses on obtaining the solution to a problem through objective linear reasoning. Critical thinking is used to analyze and critique probability, usually orally, by completing the statement "Yes, but ..." An example is "Yes that is a possibility, but I am concerned about the heat produced by electricity." The critical thinker's vertical thinking is convergent.[6] It looks for the right approach and is analytical and sequential. There is a time and a place for a little critical thinking along the path to innovation, but because this topic is readily available in the curriculum at the university level, I am focusing my book on creative thinking, which complements critical thinking. Once a person learns to be more creative, he or she will be able to generate new ideas by reorganizing, reapplying, and revising current ideas in a new way. And, TA-DA! Innovation is born.

## Why We Innovate

Innovation is the practical process of applying one's knowledge (abstractions and theories) by implementing a useful idea. That means taking the scrunched-up napkin with the sketches and following through and making the idea come to life with action.

Innovation is creating and delivering a product, process, or service of new value to the customer. The key is in identifying new value. Innovation originates in environments that support risk. Need-based innovation often comes from removing an obstacle or making new combinations of what currently exists. Opportunities to make things better and more efficient arise from human needs and from the needs of technologies. Innovations become building blocks for future innovation.

Innovation does not have to be a novel invention. It can be a new way of restructuring how we do business and how we play. The majority of innovation is implementing a new way of doing things, such as improving on old processes or existing product designs. It may be discovering new solutions to emerging problems and improving on solutions to old problems. A small part of innovation involves the discovery of pure invention that results in novel and socially useful processes, services, or objects.

If there are more iterations on existing ideas than there are novel big ideas, why is this so? Visionaries have to be in a state of dissatisfaction with the current status quo to be motivated enough to want to commit time and energy into doing something completely original. If everyone is happy and satisfied with the way things are, things stay the same. Unhappiness leads to innovation.

Passionate people who want improvement are much better to employ in innovative companies than people who are comfortable and happy with the way things are. There is a big risk in standing still and not moving forward. Status quo sets in at that point. At the same time, innovative people must respect the current business and make sure that it is not destroyed by only focusing on innovation breakthroughs.

Another group of unhappy people are the lead users of your product, who buy your product because it comes fairly close to what they really wish was invented. I was a beta tester for Adobe Illustrator before the 1988 release, and as far as I was concerned, it beat the pants off MacDraw, but the beta version had its drawbacks. Not being able to edit in preview mode was the number one concern I reported to the software engineers. They said it was an easy fix, and they revised it. It is funny how software engineers can develop a marvelous software, but unless it is used as a practitioner's tool on a daily basis, you really can not see the value of the obvious.

When Adobe Photoshop first was introduced in the marketplace, it was a photographer's software. My number one concern was not being able to save layers. That feature was fixed in the next generation of the product. At the time, I was the Art Director for *The Phoenix Gazette*, and I started experimenting with how the software could be used as an illustrator's tool. I scanned in 3D objects, actual textures, and photos that I would later manipulate. Some artists called it cheating. I called it experimentation and pushing the limits. I took the heat at my company for "playing" and "copying." It was worth it, because the next generation of the software product was marketed also as an illustrator's tool. It was a new application and a new marketing niche for Adobe. It added millions of dollars in revenue. I like to think that I was a small part of that.

Visionaries and inventive users in a state of dissatisfaction are not the only unhappy people. The most dissatisfied are people who are not purchasing your product, because it is not delivering what they want. Instead of interviewing and researching only your customer base, talk to the disenchanted. Find out why and what would tempt them to buy from you, and you may have formed your vision for the next five years.

Bill Peterson, a colleague, tells the story of the danger of asking people what they want. He said that years ago people asked for bigger floppy disks when they wanted more data storage space. In practice, disks got smaller, and today most disks are a digital museum artifact. Be reminded, however, of a quote from Henry Ford: "If I'd asked people what they wanted, they would have asked for a faster horse."

## The Economy and Innovation

The United States' economy has been in a tailspin for a number of years. The standard of living has decreased because of the loss of jobs, the housing market, and the rising cost of gasoline and food. The economy is just now starting to heal. The United States is faced with the challenge of remaining a competitive global player. For our country to maintain the competitive edge, the U.S. government is focusing on invigorating engineering and technology, and companies are hiring and retaining employees who are able to think creatively, because this leads to innovation. This characteristic has been lacking in new university graduates in the United States for many years.[7]

The majority of an engineer's and technologist's technical resources are accessible on the computer and the Internet. Many technical skill sets have been outsourced. So what will make the future engineer and technologist in the United States valuable? It will be an individual's ability to innovate.

Historically, the strength of Americans is the ability to innovate, but in the past 15 years, we have lost our lead in the world. According to the Information Technology and Innovation Foundation (ITIF) and the Boston Consulting Group, the United States ranks sixth and eighth, respectively, in patents, research spending, and venture funding. Out of 40 countries examined by the ITIF, the United States was placed last in the category measuring how much a country has improved its innovation capacity from 1999 to 2009.[8] This category factored in education, basic research funding, and corporate tax policies.

Innovation breakthroughs require the symbiotic relationship between exceptional research universities and an inspired private sector that values research, with help from government funding. Even if you do not agree with the role of the government fostering innovation, the reality is that government funding is what fueled technological innovation in the 20th century.

Let us examine the microchip, which was invented in 1958 by Texas Instruments. The federal government (i.e., NASA) bought every microchip that they could get their hands on from the manufacturers, so that within a couple of years, the price of the Apollo microchip fell from $1,000 per unit to $20–$30 per unit. And then there is the Internet, which was initially funded by the Defense Advanced Research Projects Agency (DARPA), the Defense Department's venture capital arm.

The fastest growing economies in other countries are establishing relationships with their governments to grow new innovation. Who would have guessed that South Korea would evolve into a major steel-and-iron shipbuilding industry?

America needs to revive, rebuild, and reform through changes in education, training, infrastructure, industries, and innovation models. We need to focus on science, technology, and innovation policy, which play a critical role in the wealth of a nation.

## The Need for Training and Education

Innovation skills are intrinsically tied to the economy. According to a 2004 report by the U.S. National Innovation Initiative, commissioned by the Council on Competitiveness, the United States is not providing industry leaders and university students with the skills they need to innovate.[9] The report recommends that U.S. universities teach these skills to industry leaders, educators, and university students. We, as Americans, are losing the competitive edge.

Not since the Soviet Union sent Sputnik into space has the United States been called to action to become more competitive through creative and innovative problem solving. The National Innovation Initiative Panel that studied innovation as an integral variable of the 21st century global economy was comprised of government officials, corporate leaders, and university think tank scholars—people from different areas of expertise working together.

Here we are a decade later, and have things changed? We know that the payoffs of investing in new innovation are significant. Case studies of 108 new business launches resulted in critical payoffs. Firms investing in new innovations were found to have payoffs of 38% of total revenues and 61% of total profits.[10]

For universities to teach creative thinking and innovation skills to industry leaders, these skills need to be engrained in the engineering and technology curriculum at research universities. What is also missing is how to apply these identified skills in real-world problems in industry, how to build an organization that is receptive to innovative and creative problem solving, and how to teach and train individuals to think differently.

## Food for Thought

The infrastructure of the economy changes as innovation changes. The economy adapts as a consequence of innovation. It is a constant rebirthing—an intrinsic relationship between innovation and the economy. Humans need challenges, a meaning to life, and most prefer to live in harmony with nature. It is about specialized expertise and the ability to adapt. It is not only about the money. However, the step of moving a product to market is about the money.

Engineers and technologists invent new products, improve the usefulness of objects, and use product design or technical processes to solve problems, making things better, faster, and more cost effective. With all that is happening in the world today, the value of a future engineer and technologist in the United States will be based upon his or her ability to innovate. This means that the ability to think creatively is paramount.

Engineering ideas into innovations is a social event. Products, processes, and services may be repurposed and may have unintended consequences. That is the surprise factor—the unexpected gift—in the innovative process.

In the innovation cycle, there is the initial hype phase, which is full of enthusiasm and hope. Then there is the "How are we going to apply this idea?" phase, followed sometimes by the "What the hell are we doing?" phase.

The following chapters present the key factors of learning about idea engineering, such as exploring how successful companies have challenged the roadblocks that prevent innovation, defining models of innovation, understanding how to define the problem that really needs to be solved, how to think creatively, and how to apply that knowledge in a useful way. Then finally, worldviews are presented.

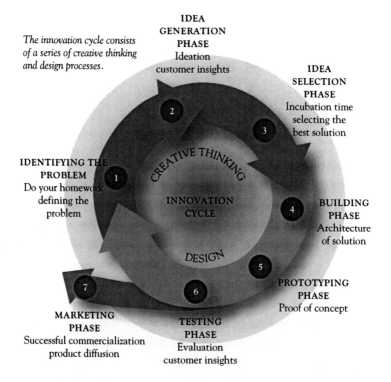

# CHAPTER 4

# Organizational Roadblocks

*Because, you know, resilience—if you think of it in terms of the Gold Rush, then you'd be pretty depressed right now because the last nugget of gold would be gone. But the good thing is, with innovation, there isn't a last nugget. Every new thing creates two new questions and two new opportunities.*

—Jeff Bezos

People are prevented from thinking creatively, because of roadblocks they have formed in their own minds. We discussed this in Chapter 1 "Getting Your Mind Right." The same thing holds true for organizations. Sometimes, the roadblocks come from working in conditions that are not optimal for thinking like a designer, inventor, or entrepreneur.

This chapter focuses on challenges that are found in places of employment that struggle with creative thinking and innovation. I identify and discuss seven roadblocks: (1) lack of visionary leadership, (2) fear of external global partners, (3) lack of resources, (4) poor internal communication, (5) organizational fear of failure, (6) running a company with no reward system, and (7) resistance to change or to the idea.

Being able to identify and understand organizational roadblocks is important knowledge for engineers and technologists. This helps capture the big picture of a company and helps to decide if it is a good fit.

Some day you will be in a position of leadership and will help build a healthy foundation for inventive problem solving. What kind of leader will you be? What type of people will you hire? What type of organization will you build?

## Organizational Roadblocks

A roadblock is a barrier that prevents further progress toward a goal. Roadblocks that prevent creative thinking in a company lead to restrained thinking and jeopardize the creative process that leads to innovation. Organizations that have difficulty with creative thinking typically lack in a shared vision, fear external global partners, have heavy workforce workloads with limited resources, have insufficient communication, fear failure, reward crisis management rather than crisis prevention, and resist change.

## Roadblock #1: Lack of Visionary Leadership

After examining case studies of innovative companies, the most important challenge to lack of visionary leadership is the hiring of the right people. The hiring of the right people in turn sets up an environment conducive to creative thinking and balances creativity and efficiency.

Challenges to this roadblock begin at the top—the hiring of the CEO, who takes on the role of the "innovation champion" and who inspires every senior manager in the company to commit to the vision. This allows the creative thinking process to take root, resulting in visionary leadership at every level. Companies are successful at being innovative when the organizational culture is receptive to creativity, and company executives take on the responsibility of shaping corporate culture.

A truly visionary leader is able to predict future needs and opportunities before they happen. He or she is a technical champion who knows how to get things done informally, understands the working dynamics of the organization, is not afraid to put his or her career on the line once in a while, is passionate about the projects and what is being done, and can deal with the constraints of bureaucracy. This person's charismatic personality inspires loyalty.

Take Marissa Mayer for example. Mayer, who was Google's first female engineer in 1999 and in 2013 is the CEO of Yahoo at the age of 37. She says that it is essential for a company leader to "act like a venture capitalist" if he or she wants to be successful. Being willing to take risks and defy stereotypes is what has helped propel her to her leadership role. And what is it like today being the youngest woman who is a CEO of a

Fortune 500 company? "Passion is a gender-neutralizing force," Mayer told a CNN reporter.[1]

One of the most significant decisions in an organization is to hire the right people all the way from the top to the bottom, who can work as part of a team. It is not just the correct selection of the top executives that makes a company innovative. Team leaders should be champions of the vision. Go a step further and also select the transformative employees—the workforce staff who are willing to solve problems.

A company cannot be successful without the contributions and talents of each employee. The right employee is a person who has a healthy fit in your particular organization. The right employee shares the vision of the company, appreciates others, and respects them. Hiring the right person for the right position is important. Hiring employees who are more talented and more brilliant than you are is a testament to your visionary leadership and allows transformation to take place.

> My model for business is the Beatles. They were four guys who kept each other's kind of negative tendencies in check. They balanced each other, and the total was greater than the sum of the parts. That's how I see business: Great things in business are never done by one person. They're done by a team of people.
> 
> —Steve Jobs[2]

It is the people who are central to the creative process. Creative thinking and innovation can be nurtured by believing in people and expecting success. This empowers employees, opens up communication, gives people freedom, and accommodates their needs. This is what visionary leaders at all levels do.

In many companies, there is an internal struggle between efficiency and creativity; so, a balance needs to be reached. Apple has been able to reach this balance quite well by approaching creativity and efficiency as symbiotic entities—an interdependent relationship. The successful creative process makes the entire company more efficient. Steve Jobs, CEO of Apple, focused first on making what he termed "insanely great" products.[3] Apple's primary philosophy about innovation is to have the vision of making great products the priority and secondarily to make a profit.

That is very much the philosophy of doing in life what you love, and the money will follow.

The goal of Hewlett Packard's (HP's) Innovation Program Office is the business impact—getting the product into the hands of the customer. They understand that it cannot be done without creative thinking; so Phil McKinney, the Chief Technology Officer (CTO) of HP's Personal Systems Group, ran an incubation program to analyze and experiment with solutions to problems.[4]

Being open to associating unrelated ideas and visualizing new relationships between concepts sets a creative environment. Intel likes to establish an environment that gives researchers both freedom and direction. Other innovative companies, such as 3M, have a deliberate creative environment, which enabled the invention of the Post-it note. In spite of the fact that this particular invention was labeled accidental, it could not have happened without the proper creative setting. 3M believes that if people are to be innovative, the one thing visionary leaders must not do is micromanage employees.

Another way to unblock the roadblock is to follow 3M's "15% rule" in which engineers and technologists spend 15% of their company time initiating and developing their own pet projects. This rule resulted in over 60,000 products in 3M by the early 1990s.[5]

When I was a professor at Arizona State University, I had the opportunity to go to dinner with several colleagues and meet with Google's leadership. One of the topics we discussed was the 3M rule, which they also adopted in their organizational culture. They allocated Friday for pet projects and made it more of a "20% rule," where employees dropped their Google responsibilities and spent the time creative-thinking and innovating. This is how many of the apps were developed, as well as the graphic play of the Google logo on the search page. This is a way of rewarding employees and giving them the freedom to be innovative.

## Challenges to Roadblock #2: Fear of External Global Partners

General Motors (GM) and IBM paved the way for outsourcing to India and other countries. Outsourcing should not be something that

American workers fear; however, outsourcing is often perceived as a practice of cutting American operations in lieu of less expensive labor in other countries. Many American workers fear having their company collaborating with external global partners. They fear that they will lose their jobs, because their jobs will be going overseas. How can this roadblock be turned around?

IDEO, a design and innovation consulting firm, has been called "Imagination's Playground" by *Wall Street Journal*.[6] They have been very successful bringing services and products to market by partnering with clients such as NASA, Nike, Polaroid, Prada, BMW, 3M, Cisco Systems, Eli Lily, HP, Procter & Gamble, Pepsi Cola, and so on. IDEO says that most companies outsource because they lack momentum, aptitude, expertise, and innovation, with lack of innovation being the top reason. IDEO believes that if a company has strength in many areas but "true excellence in a few areas," they will be successful.

Innovative companies such as Apple, HP, and Google address this roadblock by practicing selective outsourcing and collaboration. Steve Jobs depended on outsourced design manufacturers (ODMs) for Apple product manufacturing, but unlike many of his competitors, the key design decisions are not outsourced. He kept the software engineers, the industrial designers, and the mechanical engineers in-house in California. Yes, it is more costly initially, but the cost savings of outsourcing design decisions are not worth what is lost in the creative process.[7]

HP reduces operating costs associated with technology management by taking a collaborative approach to global outsourcing. HP spends less time on everyday information technology (IT) operation issues by outsourcing them. This allows the company to concentrate on innovative projects and core competencies and save money.[8]

HP also addresses this roadblock through internal collaboration. Critiques are open to all at different levels. For an organization to follow this model, it would have to undergo a cultural transformation by encouraging collaboration and form cross-functional, interdisciplinary teams.

A number of years ago, Google outsourced their billing, collections, and credit evaluations to webmasters, individuals, and other corporations. This enabled them to allocate their resources to what they do best. Vint Cerf, the chief Internet evangelist at Google since 2005, says that Google collaborates using the capabilities of the Internet, and that has led the organization to breakthrough innovation.[9] The answer to challenging the fear of external global partners is to embrace scientific research, business, and education on a global scale and use it to the company's advantage.

## Challenges to Roadblock #3: Lack of Resources

A resource is defined as a productive or economic factor required to get things done. Organizational roadblocks, such as lack of resources, limit a company's ability to initiate design and apply new value-added ideas. Limitations may come in the form of time, space, human capital, and money.

### *Time*

Time is money. Time usage is an irreversible economic opportunity. The scarcity of time is not having the resources to do the job optimally. This is followed by trade-offs and economic decisions to be made. Often, this means that more efficient ways of getting the job done need to be realized.

### *Space*

Project space to work collaboratively and creatively is probably the most neglected resource. Space can be a permanent working area or a temporary meeting area. It may mean that there is not enough physical space to comfortably support a process. Having insufficient working space may also encompass not having enough digital space to store data.

Physical space can also be used as a strategic resource reflecting the company's priorities. Who has the biggest office with the window is a corporate game that is played in some companies. The space given to a department is an indication of importance.

### *Human Capital*

Human capital is the stock of an employee's skill set that adds to the labor and economy of the company. A shortage of human capital as an innovation resource may be a deliberate decision made by company executives.

Some companies may not have employees dedicated to work on innovation, because not enough people are hired. Others may choose not to have employees other than R&D who have the responsibility of innovation, taking away opportunities for others to input new ideas.

Look closely at those organizational cultures not fostering creativity, and you will see that the entire responsibility of innovation is given to R&D and taken out of the hands of the other employees. This is a key piece of knowledge. It is as though great ideas happen "over there" in R&D, and the rest of the workforce has nothing to do with it.

In a prior study that I conducted with industry leaders from Motorola, Intel, Boeing, and Honeywell, participants stated that oddly enough, the R&D department in many organizations needed innovation training the most.[10] Organizations like Apple believe in the importance of not cutting R&D resources; however, their culture embraces innovation at all levels.

Intel suggests that confronting the problems between R&D and manufacturing is a key to innovation success. There is always a bit of tension if workers perceive those with ideas and those who have to do the work as completely separate entities. We will discuss this concept further in the chapter "Models of Innovation."

*Money*

Money is a resource to get things done. The act of working is an economic act, which involves the generation of opportunity costs. Executives may not commit to investing monetary resources, because it is not available. Other companies, such as HP, have sufficient resources, but their policy is to remain selectively lean. In HP, the decision on whether to continue an idea to the next phase—the Go/No Go milestone—is a final decision. Once the decision is accepted, the process moves quickly to completion, because it is then that resources are dedicated to the project. The faster to "Go/No Go," the better.

## Challenges to Roadblock #4: Poor Internal Communication

Companies with poor internal communication usually lack decision-making abilities and effective creative thinking. The top way to address poor internal communication is to focus on conducting efficient and effective meetings. The more efficient and effective meetings are, the faster ideas can take action. The biggest problems with meetings are that they are not well facilitated, not well planned, and there are too many of them with no purpose. Successful companies have short meetings that focus on a clear agenda with an expected desired outcome.

Edward de Bono wrote an international bestseller *Six Thinking Hats*.[11] The Six Thinking Hats process is a technique for conducting efficient meetings. Wearing one hat at a time when considering solutions to a problem allows the problem to be approached through six different perspectives:

1. White Hat: Data-gathering mode
   In the White Hat mode, the thinker asks questions, examines factual information, discusses needs, and identifies missing information.
2. Red Hat: Intuition and emotional mode
   In the Red Hat mode, the thinker goes with his or her gut feelings without having to justify it.
3. Black Hat: Logical and negative, judgmental, and cautious mode
   In the Black Hat mode, the thinker points out logically why an idea does not work with the facts, the experience, the policy, or the system in use.
4. Yellow Hat: Logical and positive, benefits-listing mode
   In the Yellow Hat mode, the thinker logically explains why an idea will work and what benefits are offered.
5. Green Hat: Inciting, finding alternatives, and creative mode
   In the Green Hat mode, the thinker is creative and discusses what is interesting. The thinker offers proposals, provocations, and changes.
6. Blue Hat: Overview and process control mode
   In the Blue Hat mode, the thinker looks at how the thinking process is going and if another hat should be put on your head.

So what hat are you wearing today? Do you need to switch hats? I recommend the book, if you want to learn more.

Apple Computer has their own way of conducting meetings. The two types of meetings they have each week are labeled "paired design meetings"—a "let's go crazy" brainstorming meeting and a "will this actually work" production meeting. These are opposite in nature and are conducted for both designers and engineers. The "blue sky" brainstorming meeting is for discussing all possible ideas. In the production meeting, idea logistics are determined and clarified.

Then Apple conducts what they term "pony meetings." These meetings address the requests of the senior management and at the same time showcase the best ideas from the paired design meetings. Michael Lopp, senior engineering manager at Apple, believes the management's "I want a pony" wishes need to be heard, as they are the ones funding the projects. Lopp incorporates management input in the design process through the pony meeting. The hope is that the paired design meetings and the pony meetings develop into a new Apple innovation.[12]

## Challenges to Roadblock #5: Organizational Fear of Failure

One of the major roadblocks to creative thinking is fear of failure and the anxiety that accompanies it. Failure should be expected as a part of the process, and should be viewed as educational, as well as evidence of action. Establishing a risk-embracing organizational culture and teaching people how to think differently is a way to challenge this roadblock. The most successful companies are open to calculated risks and are not afraid to fail. They learn from their mistakes.

Creative breakthrough thinking occurs when people are not afraid to express their ideas, because they realize that there will be no negative repercussions. Not being afraid to make mistakes opens one up to creativity and innovation. This is easier said than done. This cannot be accomplished without a risk-embracing organizational culture.

R. W. Johnson, Jr., former CEO of Johnson & Johnson, has been known to say, "Failure is our most important product." In its 107-year history, this company has never posted a loss.[13] It is not afraid to make mistakes and has many failed ventures; however, the ones that are successful are highly successful. Failure is part of the creative process. Companies who accept their mistakes, learn from them, and move on end up being the most innovative.

When people face huge and complex problems, they habitually gravitate in their thinking to what is familiar, and in doing so, they narrow the arena of potential solutions. They miss out on the majority of possibilities. What is more surprising is that they are totally unaware of their behavior. These are habit routines of behavior that are repeated, usually without even thinking about them.

People can learn to think differently through learning creative-thinking techniques in a risk-embracing environment. I believe that because higher education is in the business of teaching people how to think, the responsibility of teaching creative thinking for future technologists, designers, and engineers should rest with universities. Ideally, learning how to engineer ideas should be a lifelong venture that can be applied to your life and your work.

## Challenges to Roadblock #6: Running a Company with No Reward System

Team incentives result in performance improvement of as much as 44%. When employees are in a reward system and are encouraged to think smarter, performance improves by 26%.[14] Longer incentive programs are even more effective. Only a minority of companies give some type of incentive to employees who offer ideas for improvement of the company products.[15] Establishing an incentive plan is a management responsibility. When employees do not receive some type of reward for their ideas, they lack the initiative and do their job and nothing more. Rewards for innovative behavior and disincentives for lack of results should be in place for all companies focused on innovation. If compensation is structured to reward ideas, it would create an organization of employees who are motivated. Rewarding ideas does not always have to be monetary. Incentives such as recognition, free time, and empowering employees to do the things they enjoy will inspire loyalty, increase morale, and increase profits. Things get done that are rewarded; so, it is important that management makes sure they are truly rewarding the behavior they want repeated.

Jack Welsh, past CEO of General Electric, overcame this roadblock by putting a compensation plan in place that meant not only "rewards for the soul," but also "rewards for the wallet." In other words, employees were rewarded with money as well as other incentives.[16] Intel also overcomes this roadblock by linking rewards to performance. More than 70% of employees can purchase stock options.[17] In addition to stock options and bonus plans, Canon offers recognition, promotions, and nonmonetary rewards, such as thank you gestures and vacations.[18]

Innovative success in organizations begins with the organization's human capital—the stock of knowledge, skills, and creativity personified in each person who works for the company. Human capital is what produces economic value through the performance of work. Choosing the right people for your organization is the key to success. But if the right people are selected, and in addition given opportunities backed up with compensation, they will be highly motivated.

## Challenges to Roadblock #7: Resistance to Change or to the Idea

The Miata automobile was developed in both Japan and America and was eventually embraced as a success throughout the world. It was not always this way. New ideas often sound completely crazy and are met with strong internal resistance that often comes from an emotional place. Sometimes the key is to get external stakeholders to put outside pressure on the organization. For example, when the idea of building the Roadster (Miata) automobile was first presented, it was not received well by Japan's Mazda company. Project leaders conducted pilot tests and were able to gather information showing that potential users were very interested in the Roadster. When the survey results were presented, there was internal support. Persistence in moving this project forward was what it took. In 2000, the Guinness Book of World Records named the Miata the best two-seater convertible in history.[19]

Innovation is about embracing new ways of thinking. It is not "pie-in-the-sky." It is about survival for a company in today's postindustrial world. Innovation that begins from the top down embraces a vision and a strategy. Bottom-up innovation is driven by organizational culture that is led by leaders who drive, motivate, and encourage creative thinking and innovation.[20]

An environment that resists change and is bureaucratic, hierarchical, and cautious is most likely not to be conducive to creative thinking. These organizations continue to sustain the status quo.

## Why People Resist Change

Change is not easy, either on a personal level or an organizational level. People resist change for a number of reasons. I have placed these areas of resistance into nine categories. (1) The risk of change is seen as greater than the risk of standing still. (2) People feel connected to what is familiar. (3) People are missing the competencies to make the change. (4) People are overwhelmed. (5) People believe the change is bad. (6) There are no models for change. (7) People want to make sure new ideas are sound before they invest in them. (8) People are wary of hidden agendas among the movers and shakers. (9) The change threatens their identity and status.

### *The Risk of Change Is Seen As Greater Than the Risk of Standing Still*

Communicating a vision is not enough for most people. I have found that the best way to combat resistance to change is to manage the risks. In the United States, we seem to be initially moved by statistics. Get the people's attention first by presenting the cold, hard facts. People will decide to buy into the idea of change on an emotional level, once the rational requirements have been met. Often, having plans for both low-risk or high-risk change helps ease the transformation.

Change in an organization does not happen until the benefit of moving forward is viewed as greater than the benefit of standing still, and the disadvantages of the status quo are greater than change.

### *People Feel Connected to What Is Familiar*

Organizations moving toward cultural change must not forget the history of the company. Leaders need to recognize that if change is going to take place, the past—how things have "always" been done—must be honored. By honoring the contributions of those who brought success to the organization in the past, a leader shows diplomacy and respect for the past stakeholders, as well as the loyal employees who have devoted their lives to the company. People are social beings and loyal to what is familiar. As a leader, it is important not to make people feel like they will be betraying someone or something if they embrace the change. I have also often witnessed people remaining in the same rotten job, situation, or relationship, because it is easier to remain with the familiar than to exert energy to make a change.

### *People Are Missing the Competencies to Make the Change*

Sadly enough, many people will not change, because they do not believe they are competent enough to change. No one will actually say this, however. They will just think it. So, problems with organizational change are affected by self-esteem issues of their employees. In some cases, a lack of competence is a real problem for an individual, but in most cases incorporating effective communication and an effective training program is all that is needed. There are techniques to developing inventive thinking, which then lead to problem solving in industry. If organizations do not commit to investing in "thinking training," the development of these skill sets, and in innovation coaching, change is not likely to occur.

In my industrial experience, when an organization made the change from analog to digital (physically laying out the paper by hand to using computers), many people had to be retrained to do the very job they were initially hired to do many years ago. Some retired. Some quit. Some hung in there and got retrained. If a person does not have the competencies to make the change or does not want to commit to retraining, then it is best for him or her to make a change to a new organization.

### *People Are Overwhelmed*

Change often produces fear and even anger in people and consequently causes them to feel fatigued and overwhelmed. Apathy comes with being overwhelmed by too much too quickly. Fatigue prevents people from implementing change, even if they believe in the new idea. People who have to live with change feel fatigued when they do not feel valued. Rewarding employees during this stressful transition, and offering a way of venting concerns without getting bogged down with negativity, is important in addressing resistance to change.

## *People Believe the Change Is Bad*

Skeptics are valuable, because their voices may result in vetting a process or idea and making it better. Often when people resist change, it is good, because it can act as a devil's advocate in determining the best solutions. Invite the skeptic in the process. Do not ignore people or devalue them when they have objections to change. To do so is disrespectful and may increase resistance to change. Many of the objections may be rational and not emotional. Hmmm. Maybe you could be wrong…

I have witnessed times when a great idea for change was not received well, because the decision-makers did not have all the information to make a sound decision. Sometimes when people believe that the change is a bad idea, they are correct. When I worked in industry, I remember going through too much change too often. Before one calendar year was over, which was before the change could be evaluated for its effectiveness, another major shift in the process took place. Then a year later… another major change. This confuses people and makes them weary. There is no benefit in implementing change for change's sake. It is a bad idea.

## *There Are No Role Models for Change*

As the saying goes, "Seeing is believing." Set up pilot programs to work out any concerns on how the new innovation process works. Pilot studies are preliminary experiments used to evaluate design, feasibility, cost, time, obstacles, and user response. Check out the chapter "Designing & Testing an Idea" in this book. Remember that demonstration helps overcome resistance to change.

## People Want to Make Sure New Ideas Are Sound Before They Invest in Them

Employees need to be eased into change. They need to believe that the ideas affecting their work life have been well thought out and comprehensive. Begin with broad information explaining why changes are taking place and what the next steps will be. Outlining the internal communication process for feedback and questions is important for people to commit to a cultural change. When a visionary leader changes how things are done, people need to know specifically what is expected of them in their new roles and how they will be a part of the vision.

Honoring the accomplishments of employees along the way and reminding employees of the benefits of the change are a part of the continuous process of moving forward. Redefining organizational goals and objectives helps to clear up potential conflicts and misunderstandings. Being open with information, providing communication outlets, and not over-reacting to any negative reactions shows your good faith as a leader and your commitment to the greater good of the company. If change means a reduction in force or employees being reassigned to other roles, be upfront about it and provide outplacement options and training. Stay focused on moving forward.

When I was researching *Dancing with Dragons: Social construction of technology during times of resource stress*, I interviewed many technologists about technological and organizational change.[21] Organizations that successfully introduced new technology in the workplace included the technologists in the decision-making process. Because the workforce employees were privy to the details of the change and had a vote in the process, they bought into moving forward.

Resistance to change is found in organizations ranging from corporate America to institutions of higher education. Years ago, I presented at an academic conference called the National Collegiate Inventors and Innovators Alliance (NCIIA).[22] I met an engineer named Professor Drew*[23] from a university robotics center. He had just created an introductory undergraduate course in inventive problem solving in engineering. The course was based on stimulating innovative ideas by using both sides of the brain. It was geared toward undergraduates in science, math, engineering, and technology. We chatted for quite a while, and he told me about his struggle to bring creative thinking and innovation to the university level. He said that in the beginning, his colleagues used to look at the items in his classroom and ask why university-level students were playing with toys. He said that his laboratory and research was devalued by his colleagues, until the grant money started rolling in. Then attitudes changed. "Follow the money," my doctoral mentors Dr. Sheila Slaughter and Dr. Gary Rhoades of the University of Arizona used to say, and the power will be found. Amazing. Often, great ideas are not valued until they are backed by funding. Then the ideas become valued and sound in the eyes of others.

## *People Are Wary of Hidden Agendas Among the Movers and Shakers*

People often do not trust the motives of the movers and shakers—the leaders. Reformers are often suspected of wanting to increase their power in the organization or achieve personal glory. People often believe there are ulterior motives behind moving forward (like laying off people).

A word to the movers and shakers: All it takes is a personal and honest re-examination of motives by the leaders to understand why they are really making the change. Leaders must be open with information (both good and bad), open with communication, act in good faith, and want change for all the right reasons, or their ideas will be resisted.

The people in one organization I researched were very skeptical of the ability of the decision-makers to implement technological change, because none of the technology experts were included in the software and hardware decisions. The introduction of new hardware and software would require behavioral and skill modifications in the way work was to be done by the technologists. Because none of the technology experts were included in the decision-making, apathy, resistance to change, resentment, and discontent prevailed. In addition, no training was provided for the technology experts to successfully make the change.[24]

### The Change Threatens Their Identity and Status

The time spent on organizational change is usually longer than one anticipates. Because of the emotional turmoil involved in implementing organizational change, individuals often focus on their own agenda and interests as a survival mechanism, rather than those of the organization as a whole. This results not only in resistance to change but also in political behavior of manipulation, coercion, and concealed attempts to influence others, particularly by those who had formal or informal power in the organization.

Understand that change often affects the intrinsic rewards that made people interested in their job in the first place. If an employee cannot embrace the new rewards, then the employee must move on to another job. It is often best for everyone.

### Resistance to Change Case

In the early 1980s, I was an art director for an advertising agency. Graphic designers used wax, rub-off typography, rubylith, and ink to paste up advertisements. Several artists wondered why I was wasting my time learning how to code typography on a machine. "Within five years you will be designing your entire advertisement on the computer," I said. One man rolled his eyes and said, "That will be the day." Well, the day came sooner than he bargained.

In the mid-1980s, I was part of an innovative team from the largest newspaper corporation in the state of Arizona. We were introducing two-way interactive digital information in public kiosks, because computers were not found in every home. I ran into one of the board members at an art gallery show that I organized. At the time, I was the computer graphics production manager for the interactive venture. All he could say to me was that no one in his or her right mind would retrieve information from a television or computer monitor and that the entire project was ridiculous. When our team presented the idea to the newspaper staff, one of the editors told me that real information is found only on a printed page of the newspaper, not on bits and bytes on a tube. The concepts of interactivity, database design, and information retrieval were a prelude to the Internet, and we were actually testing the usability of the interface design before the Internet became a household word. I wonder what those doubting Thomases[25] are thinking today.

Think about this: When organizations and people become innovative, change occurs. Too many changes result in chaos. Not enough change results in mediocrity.

## Food for Thought

Take a good look at the environment in which you are working or would like to work. No company is perfect, but if you want to work for an innovative company, look for one with visionary leaders, one with healthy global relationships, one with sufficient resources, one with good internal communication, one that is open to failure as a part of the process, one with a reward system, and one that is open to change.

Innovative companies that embrace an environment of creative thinking and innovation allow employees to pursue ideas, fail, and share information. They make use of trained professionals and pursue collaboration with trained talent from research universities. They have resources and buy-in for the creative thinking process to lead to the ultimate goal of innovation.

Working in an environment of creative thinking does not mean being surrounded by chaos. Creative thinking is a serious process that can be systematic. Serious creativity[26] does not just aid in finding solutions to problems. The process itself changes the people in an organization. It is transformational. It changes how people think and how people approach work.

Nonvisionary companies are stagnant, bureaucratic, autocratic, commit to policies that discourage entrepreneurship, and have no incentives for creative thinking. It is very difficult to be innovative in this environment.

The bureaucratic environments of large organizations are not always conducive to out-of-the-box thinking; however, many large companies, as well as smaller companies, have been successful innovators for our society. They need to continue to successfully challenge these idea roadblocks to contribute to the sustainable economic growth of American businesses in the 21st century.

# CHAPTER 5

# Models of Innovation

*We've made a lot of mistakes. And we've been very lucky at times. Some of our products are things you might say we've just stumbled on. But you can't stumble if you're not in motion.*

—Richard Carlton, 3M

The challenge that traditional large business organizations have is that they are built for operations and effectiveness, not for innovation. This by and large is not because the employees do not have the capability of being creative thinkers and innovators and not because the technology is not available. This has to do with the strategic limitations of the leadership and problems moving ideas to market.

In this chapter, I summarize the evolution of American business. I introduce four business models of innovation and seven idea team models that I named and categorized based on my research on innovative organizations. These have proven to be successful by companies using these models.

This knowledge is significant for engineers and technologists, because understanding how companies organize decision-making will make you better members of the team and better leaders. What model is your company using? When you lead, what model do you choose? Will you invent another model?

## Business Evolution

Management has evolved in America over the last 60 years. In the 20th century, business schools perceived their human capital (the people working in the company) merely as tools to make widgets on the production assembly line. Business schools even viewed themselves as trade schools. By the middle of the 20th century, their worldview changed. Business schools began to stop perceiving a business organization as a piece of equipment, and human capital as mechanical instruments. The metaphor was replaced by a transformation through strategic thinking.

By the 1970s, senior executives realized the importance of strategy in business—the science and art of developing a plan to reach a goal. By the 1980s, the focus of business was the quality movement of efficiency and effectiveness.[1] Those were very lucrative years in the business world, and consequently there was resistance to change, because businesses were sustaining success.

In the past, times of change in organizations were followed by times of stability, allowing the organizations to assimilate the change until the next change was introduced. Because of the introduction of technological change, times of stability have gotten shorter and shorter. By the 1990s, the business world was forced to be open to change. Business executives moved their companies forward and innovation initiatives became the new strategy.

The late 20th-century business school viewed itself as an academic research institution.[2] Ironically, there was a time that business organizations had no use for PhDs. University researchers were perceived as only useful for academic institutions. But in the late 20th century, business organizations started hiring PhDs to do research and development (R&D), because it was more advantageous to have them at their disposal full time than to hire them as consultants. Companies saw the value in research for business.

Just because innovation in organizations is 1% inspiration and 99% perspiration, it does not mean a person cannot have fun while inspired and perspiring. In fact, "productive fun" has always helped me enjoy work. If someone is going to spend at least eight hours a day working, it should be fulfilling and rewarding.

## Four Business Models of Innovation

My research has led me to categorize successful and innovative companies in four desired business models of innovation: (1) ideas + leaders, (2) ideas + processes, (3) ideas + motivation, and (4) ideas + leaders + teams + plans.

## Model #1: Ideas + Leaders

INNOVATION MODEL #1

It is difficult, but not impossible, to create change in a large organization. Peter L. Harris, along with his great ideas, is the epitome of Model #1: Ideas + Leaders. When my son Rocky Harris worked at the San Francisco 49ers in the media area, he had the opportunity to interview Peter Harris, President and CEO. Rocky told me that this charismatic and candid man approached running a National Football League (NFL) team just like he did running any of his other business ventures. Peter Harris was known for turning companies around with his playful and innovative ideas.

Peter Harris began his career stocking shelves at Gemco in California. Thirteen years later, he became the CEO of the company. These years were the peak of sales for Gemco. In the mid 1980s, he moved on to become an investor partner with FAO Schwartz, which became a leading toy store under his leadership. Tom Hanks' movie "Big" from 1988 was inspired by the company and also by Peter Harris. He was the master of making work fun. Peter Harris then moved on to work at Accolade, a video game company, and then to the San Francisco 49ers in 2000. In the four years he worked at the Niners, he changed the business culture from instability to a common vision. Today, he works as a consultant in crisis management.[3]

Creative thinking and innovation is not just about producing products. It is about how one leads, how one listens, and how one treats employees and customers. A past innovation leader, who illustrated all these characteristics, was Walt Disney. He believed a company's top resource was their people. He was quoted as saying, "I happen to be kind of an inquisitive guy, and when I see things I don't like, I start thinking, 'Why do they have to be like this and how can I improve them?' " Today, the Walt Disney Company's culture of internal collaboration leads to a plethora of ideas, because all its employees (who are called "Cast Members") act in leadership roles, commit to unconventional thinking, and are not afraid of successful failures, according to Bruce Jones, Programming Director of The Disney Institute.[4] When brainstorming for ideas, the words out of their mouth are "Yes, and ..." instead of "Yes, but...." The company culture is based on the three mantras: (1) Everyone can be creative, (2) ideas do not belong to people, and (3) responsibility for success belongs to everyone. Under Walt Disney's leadership, simplicity, global worldview, and success measurement are the core values for the Disney culture. Performance is rewarded, feedback and ideas flow freely, and opportunities for training and coaching are available.

The model of innovation combining ideas plus leaders is also seen in Steve Jobs and Apple, Inc. Leadership is not always enough, because one leader cannot fight the system alone. The pressure of short-term results is put on the shoulders of the leader. Jobs, best known as the leader of Apple, Inc. left the company in 1985 after a power struggle with the board. Eleven years later (1996), Jobs returned to Apple as an advisor and then CEO. Apple went from almost filing for bankruptcy to being profitable by 1998. This milestone is regarded as one of the greatest company turnarounds in history.[5] It took the perseverance of a real leader to do this. The best leaders set up their companies for success. When they leave, the companies will sustain results even in their absence. Steve Jobs has passed along a company of innovation that will produce results for years to come.

## Model #2: Ideas + Processes

INNOVATION MODEL #2

Model #2: Ideas + Processes is a model for companies relying on a progression of steps to develop a product. A well-known company that uses this model is Deere & Company, which is commonly known by its brand name John Deere. It traces its origins back to Grand Detour, Illinois, as a blacksmith shop, where in 1837 John Deere produced steel plows for farmers living in his town. Today, it boasts a presence in over 30 countries.[6]

Deere & Company's business model for innovation is a combination of ideas and processes. This business has mastered developing processes with sequential steps that can be used repeatedly throughout the company. The success of this business model relies on each initiative being similar to the prior process. The standard processes of yesterday have grown to globally standard processes.

An example of an innovative product that was incorporated into the assembly-line process in the 1990s at John Deere is Optiflex, which is a process design software.[7] At the time, it moved the company from pen and pencil to computer. The use of the software initiated a process flow, which prevented having excess inventory.

The John Deere assembly line was complex; so, assembly-line workers were organized into 12 modules with a leader at each phase of the process. Because the pay of the frontline workers depended on both the quantity and quality of the planting equipment produced, they had a vested interest in the success of the process. The Optiflex software provided the free flow of information from quality control to assembly schedules. This resulted in the distribution of authority and an improved manufacturing process.

Another well-known company that uses the Ideas + Processes model is Amazon, which is a multinational e-commerce business that began in 1994 as an online bookstore. Amazon soon diversified into selling CDs, DVDs, software, clothing, furniture, jewelry, toys, streaming movies, and so on. It was founded by Jeffrey Bezos in 1994 with headquarters in Seattle, Washington. It survived the dot-com bubble burst and turned its first profit in 2001.[8]

Amazon has been innovative through constantly reinventing its business model—a process improvement system, based on Japan's Toyota Kata.[9] The premise is that not finding any problems in the process becomes the problem. Problems are looked upon as opportunities for making changes to improve the process. Problems are what keep a company in motion and continuously innovating. Recognizing a problem is how we learn. It is not personal. A problem may not be what we want, but it is not good or bad. Understanding a problem is not the same as accepting a problem. A problem is blamed on the system, not the people; however, this does not discount the importance of how people deal with the problem. Problems are not a threat to be hidden or quickly solved by the first idea. Take the time to observe the specific step in the process until the cause of the problem becomes apparent. Addressing the thousands of small failures along the process system is what has led to Amazon's success.

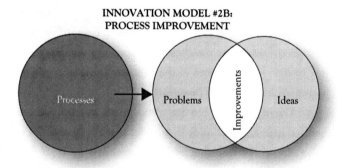

INNOVATION MODEL #2B:
PROCESS IMPROVEMENT

This "Ideas + Processes = Innovation" model at Amazon has been tweaked into a process improvement model using the Plan-Do-Check-Act (PDCA). The PDCA cycle is a scientific experiment, which gives Amazon employees a practical way of getting information at every stage of the process before it becomes a big problem. Any problems in the process are identified, which lead to ideas on process improvement. There is speculation that the PDCA process may have been introduced to the Japanese in the 1950s through a lecture course by William Edwards Deming using Walter A. Shewhart's circular process steps.[10]

Simply stated, the PDCA process is not just a technique. It is a method of managing and thinking. Instead of saying, "Let's see if this will work," ask, "What do we have to do to make it work?"

## Model #3: *Ideas + Motivation*

INNOVATION MODEL #3

Companies who believe in the power of the frontline employee are examples of Model #3: Ideas + Motivation. Nucor, the largest steel company in the United States, began its role as a start-up company with CEO F. Kenneth Iverson as their creative thinker and leader. Under Iverson, Nucor became known as one of the most innovative industrial manufacturers. Iverson believed that employees would make that extra effort, if they were self-managed, rewarded monetarily, respected, and empowered to make decisions. In the 1980s, he paid workers based on individual performance. He adopted an owner–operator's mentality for his frontline employees, instead of insisting they wait for instructions from superiors.[11]

This way of thinking has empowered workers by flattening the hierarchy, and it moved the company at the first half of the 2000 decade to profit margins surpassing eBay, Amazon.com, and Starbucks.[12] The Nucor culture embraces taking risks on employees' ideas and accepts intermittent failure. In fact, it has gained recognition for printing every employee's name on the cover of the company's annual report.

Nucor's model for innovation is a combination of growing ideas plus empowering employees through motivation. Ideas are sparked by cross-training employees and rotating employees among plants. Employees are motivated by opportunities to receive generous bonuses for innovative ideas that improve the efficacy of the production process.

## Model #4: Ideas + Leaders + Teams + Plans

INNOVATION MODEL #4

Combining ideas plus leaders plus teams plus plans is what Model #4 is about. One of the best representatives of this business innovation model is Nike. Nike, an American Fortune 500 company, is one of the world's top suppliers of footwear and apparel. Nike was awarded the title of "the number one most innovative company of 2013" by Fast Company with profits up 57%. Nike's 2012 annual revenue topped $24 billion in its fiscal year ending May 31, 2012, which was an increase of 60% since 2006, when Mark Parker was named CEO.[13]

Nike employees are not afraid to embrace adversity. In fact, complete disruption in effort appears to be one of its secrets to great ideas. If someone walked into the Nike testing laboratory, known by insiders as the "Innovation Kitchen," he or she might find that it is full of swoosh scientists and engineers and a Nike trainer with embedded sensors that measure the pressure exerted when the shoe hits the ground. The trainer might be jumping or sprinting in place. The scientist yells "What if ...?" questions like "What if you jump side to side with one leg off the ground?" and "What if you run backward?"

The data is collected and transferred wirelessly to an iPhone or transferred to an iPhone connected to a Macintosh computer. The track-and-field avatar on the screen is actually the Nike trainer. The screen on the Mac looks vaguely similar to those 1980s video game systems like Nintendo.

Nike employees view innovation as a process, not an end result. Nike had two remarkable innovations in one year—the Flyknit Racer and the FuelBand.[14]

Flyknit Racer is a recent innovation. They are shoes so light that they feel like wearing a sock. The entire manufacturing process had to be completely transformed to produce this new product.

Nike is not only willing to transform its manufacturing process, but it is also willing to redefine who they are. Nike is undergoing a digital transformation and leaving its footprint in the world of bits and bytes. Take, for example, their other innovation called "FuelBand," which is a wristband that tracks daily physical activity. CEO Parker takes this concept a step further and imagines the "cool factor" of future body feedback using different sensory inputs. What if your heartbeat or the movement of your body could change the music in your earplug? Or if the beat of your heart could change the color of the wristband? Is this insane? Are they just goofing off and wasting time? No. That is Nike-style innovation.

Another organization known for innovation is Alcoa. It has been ranked one of the most admired metal companies. In 1888, Alcoa invented what has now become the modern-day aluminum industry with markets today in aerospace, energy, alumina, aluminum ingot, automotive building and construction, commercial transportation, consumer electronics, industrial products and services, packaging, and oil and gas. Approximately 75% of all the aluminum produced since 1888 is still being recycled today.[15]

Alcoa's impact spans across the world to over 200 locations in 30 countries. An integral part of what Alcoa delivers to its customers is sustainability. The company believes in giving back to the community by contributing to nonprofit organizations, improving the environment, and preparing tomorrow's leaders in science, technology, engineering, mathematics, and manufacturing.

Innovation is an essential part of Alcoa's business model. Its success is based on its formula of applied engineering plus research plus development. Its teams of scientists, engineers, and researchers around the world are committed to the next Alcoa idea. Scientific leaders in research centers and universities in India, Europe, Russia, China, and the United States are all globally connected to Alcoa. Collaborating with external partners in R&D, such as industry leaders, universities, and national laboratories, has helped Alcoa move new ideas to market efficiently and at a greater speed. The company invests 1% of its revenues in R&D to continue to build the foundation of innovation.

Alcoa's plan is to increase customer value through innovation. Even during single-digit end-market growth, Alcoa is able to reach increases in revenue. Because customers in each market are looking for cost savings, specifically in the areas of recycling, greater energy efficiency, and fuel reductions, Alcoa innovation enables customers to reach their savings goal.

## Seven Idea Team Models

The business model of innovation implemented in a company reflects its philosophy of creative thinking and innovation. One or more idea team models may be incorporated within each business model of innovation, as long as the teams communicate with each other.

Within each of the four business models of innovation, a variety of idea team models can successfully be used by companies. The seven idea team models I am introducing are composed of the following: external members (one team), internal members (two teams), and mixed members (four teams). Each idea team model has a different purpose and a collection of different members. The common thread is that the idea team models result in innovation being managed, structured, and/or measured for success.

### *External Idea Team #1: Thought Leaders Network Team*

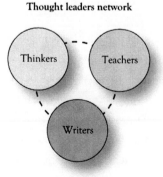

IDEA TEAM #1 (external)
Thought leaders network

Thought leaders make up a network of thinkers, teachers, and writers who inspire, originate new perspectives, and act as mentors. They have the global vision and insight to predict the next big thing. They know how to ask the right questions.

The Thought Leaders Network Team is a "sandbox" approach to an idea team. Remember when you were a child and the sandbox was a great place to play, because it was fun and creative? You sat in the sandbox with your bucket and shovel and concentrated on the building of your castle. When others joined in, they had their own ideas. Often you had to compromise, if you wanted to play with them. If they brought their own sandbox toys or liked yours better, you had to learn to share. It is no different in a grown-up work environment. This is what I call "productive fun."

The purpose of the Thought Leaders Network Team is to bring outside consultants, university researchers, and expert practitioners together as an intertwined system of external knowledge for the company. Traditionally, these thought leaders are contacted when needed by the company. Often, the thought leaders are called upon to be in a panel to discuss a particular topic or project.

Today, there are additional types of thought leaders that are actually in the business of helping a company and people be more innovative. One such business is the Stamford Innovation Center, located in Connecticut. This organization is a network of thought leaders that provides a place to gather and exchange resources, expertise, and talent. Stamford Innovation Center is a place where companies that are early start-ups, or individual entrepreneurs, can benefit. Even internships are offered.

Pitney Bowes also sees the benefit of collaborating with early start-ups and entrepreneurs in the Stamford Innovation Center. Pitney Bowes conducts an entrepreneurial competition in collaboration with the Stamford Innovation Center to explore new ideas built upon the Pitney Bowes technologies and services. So, all benefit.[16]

Another type of thought leader in today's world can be formed through the use of social media, such as blogs, LinkedIn, and Twitter, where an organization becomes the thought leader. But value for customers has to be added, because those customers become the social capital of the organization. Research has demonstrated that companies like IBM have benefited from online social relationships at $948 per contact per year, according to Jay Baer, a social media expert, host of Social Pros podcast, and founder of http://convinceandconvert.com.[17]

Deere & Company, an example of Model #2 (Ideas + Processes), has also utilized the External Team #1 Thought Leaders Network Team model. Years ago, John Deere discovered the value of online thought leaders quite by accident. In the late 1970s, John Deere tried to cut costs by scaling down the options offered to its farmers. This did not go over well with the customers. A program called Vision XXI was then launched to offer a variety of choices to its farmers without the cost of building a new factory. The assembly-line process became more complex as time went by, and the hand scheduling became a nightmare. Bill Fulkerson, a former math professor who was hired as John Deere's staff analyst, used the Internet to ask thought leaders for help with ideas to improve the automation. A computer scientist, who worked at the U.S. Department of Defense at the time, replied to Fulkerson. Several other computer scientists acted as thought leaders to help find a solution. This resulted in an assembly-line software called Optiflex being developed and installed in 1994 in six of their plants.[18]

## Internal Idea Team #2: Cross-Pollinating Senior Management Team

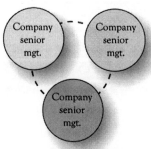

IDEA TEAM #2 (internal)
Cross-pollinating senior management

PURPOSE
- identify obstacles
- establish reward system
- form company culture
- develop processes
- measure success

The Senior Cross-Pollinating Team is composed of senior management from a variety of company departments and no external members. The purposes of this idea team are to identify obstacles that hinder innovation within the company, figure out reward systems for great ideas, work on forming the company culture into one that embraces innovation, develop processes that are conducive to innovation, and figure out ways of measuring the success of new ideas. A big task indeed; however, HP, a global leader in information and print technology, has been very successful with this model.[19]

## Internal Idea Team #3: Cross-Pollinating Expert Team/Project Team

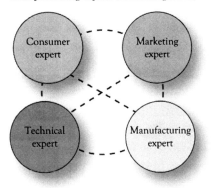

IDEA TEAM #3 (internal)
Cross-pollinating experts (Non-management)

PURPOSE
• identify opportunities for company growth in innovation

The Cross-Pollinating Expert Team or Project Team is composed of non-management experts and researchers of a company from various departments, such as marketing, manufacturing, consumer, and technology. The members meet to discuss opportunities for growth that lie across the boundaries of their departments. No external members are included.

If the Cross-Pollinating Expert Team is a team of employees assigned to work on a particular project, it is often called a Project Team. This type of idea team is found in companies such as IDEO, Apple, Google, and so on. If splitting into project teams using employees and a project leader is what a company does, I suggest employees select their first and second team choice, rather than having team leaders choose which employees they want on their teams. This allows employees the opportunity to work on projects in which they are interested. A team with talent, not just seniority, with employees from diverse backgrounds is often the most creative.

Southwest Airlines has used the Cross-Pollinating Expert Team model successfully on a much larger scale by expanding internal collaboration through the platform of a virtual collaboration team. Southwest Airline's thousands of employees nationwide take on the role of innovators by sharing novel ideas or possible improvements for the organization. Each employee brings his or her area of expertise to the table.

This team is made possible through the utilization of Spigit, an innovation management solutions and idea management software. The software is designed to streamline internal collaboration, used to identify problems that need to be solved, and generate ideas that reduce costs, enhance customer relations, and generate revenue.

Each Southwest employee has access to collaborating virtually with other employees throughout the nation. Monthly and quarterly incentives are awarded for top ideas. Angela Vargo, manager of product development at Southwest Airlines, uses this software platform to tap into the collective intelligence of the organization.[20]

At the university level, the senior project is often based on this idea team model with teams of students from multidisciplines, such as engineering, business, industrial design, computer graphics, and computing, who act as cross-pollinating experts. The difference in the team model is that the teams must present their innovative senior project ideas to the professors and industry leaders for the Go/No-Go decision.

## Mixed Idea Team #4: Senior Enterprise Team

**IDEA TEAM #4 (mixed)**
Senior enterprise team

- Company CEO
- Company executive
- External tech expert
- External marketing expert

**PURPOSE**
- drive the growth of the company through innovation

Senior Enterprise Team is made up of high-level internal members (i.e., CEO and executives) and several thought leaders who are technical experts and marketing experts (i.e., consultants, noncompetitive leaders, retired CEOs and executives, venture capitalists). The purpose of the Senior Enterprise Team is to drive the growth of the company by discovering new or improved ideas and evaluating their merit. They meet at defined times determined by the company executives—monthly, three or four times per year, and so on. The internal members decide whether an idea gets the Go or No-Go decision. Kimberly-Clark uses the Senior Enterprise Team as their idea team model with the idea team made up of high-level internal members.[21]

### Mixed Idea Team #5: Open Innovation Team

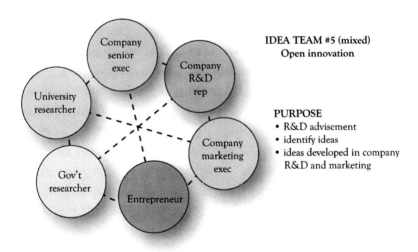

An Open Innovation Team is composed of a mixture of internal members (such as senior executives, R&D employees, marketing executives, and other stakeholders), and external members (such as university and government researchers, entrepreneurs, and laboratory experts). This idea team revises the traditional R&D department of a company by identifying ideas and using outside resources. Then, those ideas are developed in R&D and marketing. Alcoa has been successful with this idea team model.

The term "Open Innovation" was coined by Dr. Henry Chesbrough of the Haas School of Business, University of California, Berkeley.[22] In today's world, knowledge is globally available. This means that companies cannot depend only on their own research, but should also advance their growth by also obtaining intellectual property or inventions from other companies.

This often results in market-driven collaboration, as was the case with DuPont and Alcoa in 2006. The two corporate leaders joined their expertise in high-strength material science and technology to develop a construction product that provides protection against hurricane-propelled debris in hurricanes up to a Category 3 storm (wind speeds up to 130 mph). The innovation is a durable architectural panel system called Reynobond® with Kevlar®. It is essentially a safety net for buildings. DuPont's™ Kevlar® ultra-strong fabric is inserted in architectural panels made of Alcoa's Reynobond® aluminum composite material. This was a win-win situation for both companies.[23]

HP laboratories are based on the concept of open innovation. They bring global researchers, scientists, and entrepreneurs together to discover the next breakthrough innovation. HP has innovation programs in corporate and government research, along with internships and fellowships.

Phil McKinney, former CTO of HP's PC division, spent nine years at HP. During his tenure that ended in 2011, McKinney turned the Personal Systems Group around from losing $1.5 billion to being the number one PC maker with a $2 billion profit. This visionary knew how to prevent good employees from leaving the company for startups. Through his Innovation Program Office (IPO), he found and funded the ideas of HP employees.[24] McKinney valued HP human capital, and workers were entrusted to develop their creative thinking genius. When employees are included in the innovation process and expected to be creative thinkers, many will rise to the occasion and participate and innovate.

In my career, I was fortunate enough to be awarded a grant from HP that helped fund my IDEA Laboratory at Arizona State University. We used digital equipment that HP gifted to us, such as cameras, scanners, color proofers, and printers, to produce a bound book *Whispered Words: Your Story, Your Culture*. This was a book about the personal narratives of youths and their struggle to fit in society. The book was a collaborative effort and was written, illustrated, designed, and printed by the students at Arizona State University, Purdue University, and the American Indian reservations in Arizona.[25]

### Mixed Idea Team #6: Collaborative Best Practices Team

A collaborative effort between both internal and external stakeholders forms the Collaborative Best Practices Team. Their main focus is to attract new talent in the company, foster innovation, and help perpetuate the organizational culture of innovation. This is done through social networks such as Facebook, Twitter, and blogs.

Pitney Bowes is a multibillion dollar global leader delivering technology solutions and service to over millions of customers. The company provides communication, document, mail, and print lifecycle services. The journey in successful innovation began in 1902 when Arthur Pitney was awarded a patent in his first double-locking, hand-cranked, postage-stamping machine. Today, the Pitney Bowes portfolio includes over 3,500 patents globally with employees in 100 countries.[26]

Pitney Bowes collaborates with customers, external alliances, and employees on all levels. The company has a Customer Innovation Center in Connecticut for customers to test out the newest technology products.

One example of a partnership with an external alliance is when Pitney Bowes collaborated with HP to offer the IntelliJet™ Printing System, an integrated print-and-mail solution.[27] Another example is when they collaborated with McDonald's® to create a document management service that improved operational efficiency with a better-than-99% on-time and accuracy performance.[28]

Pitney Bowes understands that measuring collective intelligence[29] is difficult in some companies when middle management views it as a time waster and employees will not participate. Managers at Pitney Bowes have a different mindset, partly because offering challenges for frontline employees to tackle is a part of their job. Big ideas or incremental ones are welcome. Even ideas that result in small changes often lead to frontline problem solvers being asked to join with others to solve larger issues. These teams of people then collaborate on a global level, resulting in innovation.

An example of productive fun is when Pitney Bowes introduced an internal competition called "Innovation Idol," which was based on the television show "American Idol," to motivate the employees in the internal software business. Many of the employees collaborated with colleagues in another division of the company. The six most promising ideas were selected to make onstage presentations to the management. Murray D. Martin, the now incumbent Chairman, President, and CEO, wore a Simon Cowell mask to represent the most critical judge on the TV panel. This resulted in three contracts worth 300,000 euros ($383,940) for the employees.[30]

## Building a Dedicated Idea Team

Companies sometimes call the dedicated idea team that is successful the hot team, enterprise team, R&D team, inspiration team, or the innovation team. But in reality, the dedicated team is NOT the innovation team in an organization. I call the dedicated team the "idea team." It has the time, energy, and money to come up with novel ideas and improvements. Innovation takes place when the rest of the company partners with the dedicated team to produce a new product, service, process, or improvement.

## *A New Idea Team Model #7: Managed Partnership*

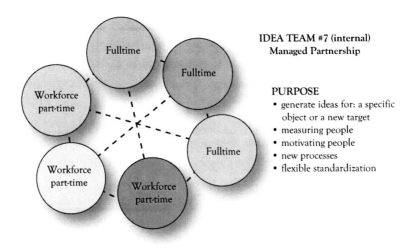

The idea team functions best when it consists of a combination of dedicated full-time idea team personnel and shared part-time members from the frontline. I call these frontline workers, who are part of the idea team, the "workforce team." These people are in the middle of building the products, applying the process, and delivering the service. If anyone is going to know where the problems exist, it is those who are directly affected. Eureka! What a concept. This results in the idea team becoming a managed partnership. The workforce team, which is the company team, is designed for efficiency. Their projects must be on budget, on time, and follow the specifications of the project.

The first item that the management needs to identify is a list of skill sets that are needed for the idea team and then hire those people either internally or externally. The mix of internal members must be carefully considered. It is a great way to offer a rewarding opportunity for a current employee, especially if the person is a creative thinker and open to change. Too many internal members spell problems of familiarity, which can lead to doing business as usual. Adding outsiders to the mix helps change the culture of the team.

Once the skills are identified and the people are hired, new job titles and descriptions must be determined, rather than using existing job titles.

Meeting in a separate location also helps the idea team feel like they have their own physical space and a distinct culture. This requires a power shift that may initially feel uncomfortable to some members. This zero-based[31] organizational model requires that all ideas start fresh and new.

The idea team needs to establish two team goals: (1) to generate ideas that enable the company to achieve specific and quantifiable objectives and (2) to generate ideas for a new target, which is independent of the strategic management. The idea team needs to address issues, determine ways of measuring objectives, motivate people, formalize new processes, and establish flexible standardization. It is not innovation until ideas are applied and working. It takes managing the ideas to bring them to fruition.

Some idea team members in this model will be full-time participants and others part-time participants who be in the workforce the rest of the time. It is important for the full-time idea team staff to respect the limited shared time of the part-time members, who divide their time between the workforce team and the idea team. The limitation of the shared members is that they must also sustain ongoing operations. Contrary to popular belief, the shared staff actually ends up doing most of the work. They are not only the liaison between the idea team and the workforce team, but once the idea goes past the final prototype stage, they are the ones who have to move the prototype to the workforce team and integrate it into the ongoing operations. Both teams are dependent on each other to take creative thinking successfully to innovation.

## Idea Teams and Workforce Teams

Start-up companies are different from companies that have been doing business for a long time. A start-up company can spend more time on innovation, because they do not have to focus on continuing operations. Having an existing business that wants to be more innovative requires a culture change of creating a work environment allowing for two potential antagonists to work together—the idea team and the workforce team. The stronger the ongoing operation, which is the workforce team and the known entity, the more of an obstacle innovation becomes to that organization.

To understand this potential conflict, we need to realize that the workforce team, which is a large and complex entity, practices that which is repetitive and that which is predictable. Most business owners want productivity and efficiency. They want the product to be better, faster, cheaper, yet reliable. The workforce team delivers growth and productivity. It follows the company plan. It has a system worldview and is driven by data analysis. Because its activities are based on proven facts, it is reliable. Commercial success leads to a focus on profit.

On the other hand, innovation thrives in an experimental kind of environment. The small and simple idea team is an unknown entity that focuses on generating ideas and improvements for a company. It has an antagonist worldview driven by hypotheses. Innovators must learn how to break some rules to be successful with nonroutine tasks with lots of uncertainty. The idea team follows a special plan, and is distinct from the company. Contrary to what some believe, creative thinkers can be accountable, manageable, and disciplined, but their mission and organization is clearly different and separate from the ongoing operations.

| IDEA TEAM | WORKFORCE TEAM |
| --- | --- |
| Antagonist worldview | System wordview |
| Driven by assumptions metrics | Driven by data analysis |
| Experimental | Based on proven facts |
| Uncertain | Predictable and reliable |
| Non-routine tasks | Routine repeatable tasks |
| Unknown entity | Known entity |
| Generates ideas & improvements | Delivers growth and productivity |
| Follows special plan | Follows company plan |
| Organized differently | Organized status quo |
| Distinct from company | Company team |
| Small and simple | Large and complex |

The truth is that both entities need each other. The idea team and the workforce team can successfully coexist in the same company, if they respect each other's contribution to the success of the business. Innovators need ongoing operations to produce a profit margin to pay for innovation. Ongoing operations need ideas, such as continuous process improvements and product development initiatives, to keep the company competitive.

## Food for Thought

It would be nice to work in a perfect model company—a positive, creative, and innovative environment where people park their egos at the door, do not have to be right all the time, have the freedom to fail, the freedom to contribute, the freedom to ask questions, and where customers love the innovations. Well, nothing is perfect. And there is no way to avoid all disparities at work, even in the most innovative organizations. That is just life in the big city. The benefits have to outweigh the differences for you. And if you can, embrace those challenges.

The companies I have mentioned have proven themselves to be innovative at some point in time. Some of the companies get higher reviews from their employees based upon the site in which they are located or upon the period of time they posted the review. Sometimes even in innovative companies, there are pockets of inherent problems in the way people are managed, inherent problems with clashing personalities, and/or lack of resources that are found in one company site and not in another. Sometimes, the problem is brought on by the country's economy. Sometimes, even in the best organizations, something goes amiss.

As with any model, it is important to note that the selected innovation model is what the company is striving to be. In a perfect world, we would have corporations who foster creative thinking and innovation and produce happy and empowered employees, satisfied customers, and inspired leaders. But because organizations are made up of imperfect human beings, people are the critical link as to whether or not the company actually does what it strives to do, treats employees and customers the way it intends, and innovates with productive fun. The key is always the people.

# CHAPTER 6

# Identifying the Problem

*If you only have a hammer, you tend to see every problem as a nail.*
—Abraham Maslow

Einstein has been credited with saying that if he had one hour to save the world, he would use 55 minutes defining the problem and five minutes finding the solution.[1] A little extreme, but it definitely demonstrates the significance of clearly defining the problem. The quality of the solution is in direct correlation with the quality of the problem statement.

When you can discriminate between what exists now and what you want that is better, a problem is identified. Intel says that many times, creative activity is wasted on trying to solve the wrong problem. Steve Jobs of Apple has been known to say that his company is successful at innovation because they focus on select problems to be solved and "say no to 1,000 things."[2]

Because the biggest mistake people make is putting a lot of energy and resources into trying to solve the wrong problem, creative activity is wasted. So, the very first thing in the process of creative thinking and innovation is to take the time to define the right problem that needs to be solved. This is done by doing one's homework, being open to learning, and embracing a deeper understanding of the unfulfilled gap between life as we know it and what we desire. This comes with being open to ideas with a systematic incubation period and experimentation.

In this chapter, I demonstrate how to deconstruct the problem to define it, and I present visual ways of defining a problem (i.e., fishbone diagram, concept and mind mapping). I show you how to construct a clear problem statement and present an example of how Amazon defines a problem.

This information is important to the engineer and technologist, because often problems are reported as symptoms. If engineers and technologists set up a systematic approach to define the real problem, it will prevent wasting time solving the wrong problem. Treat the cause, not the symptom.

## Deconstructing the Problem

There are basic questions that are used to build a foundation to define the problem. This involves deconstructing what one believes is the problem to find out the causes and consequences.

### *Question 1: What Do You Think Is the Problem?*

What does not work? What needs to be changed? What prevents you from reaching your goal? Spend time observing what is perceived as the problem to develop a deeper understanding. Team members may opt to take a field trip out of the office for inspiration or spend time directly observing how the product or service is used or not used. Observe what is missing and pay particular attention to the competitor's product or service and how it differs. What is everyone else not doing? Identify any studies that have results addressing the problem.

Try stating the problem as a reversal. Convert the positive statement into a negative one. Ask "What if?" Flip-flop the results. Here is an example of the positive statement: "We need to identify how to improve customer service." Here is an example of turning that statement into a negative one: " We need to identify how to make customer service bad." By taking the reversal approach, teams come up with more ideas, because it is a different way of thinking.

There may not be enough information to actually define the problem, so the overall problem may need to be restated in general terms at first. Each problem is a piece of a larger problem. Reword the problem using "hypernyms"—words that are more generic with a broader meaning; for example: "motorcycle" becomes "motor vehicle." This tool can help provide more information to help define the problem in broader terms.

An online source that is helpful is WordNet, which is a lexical database for the English language. It is a combination of a dictionary and a thesaurus that can also be downloaded at no cost. It was created by Princeton University and is still maintained by the Cognitive Science Laboratory.

Once the crux of the problem is figured out, reword it. Instead of saying "The time from product order to packaging is too long," ask instead "How can I decrease the time it takes for a product order to get to the packaging department?" Use the thesaurus and take single words and substitute them; for example: "increase" becomes "attract." Try changing the sentence so that it shows a benefit; for example: "increasing sales" is changed to "making your job easier." Then specify the problem by using hyponyms, which are words that are more specific; for example: "motorcycle" becomes "dirt bike."

### *Question 2: Why Is This a Problem?*

Begin by asking why this is a major problem in the first place. This will help identify consequences. Then ask, "Why else is this a problem?" until about five consequences are listed. Why is finding a solution significant?

Look at the problem through different glasses. Who is affected by this problem? How will it impact others? Is the problem yours alone? A key part of gathering information is to talk to the people who are affected by the changes and give them a chance to provide input. A marketing survey is another option to gather some of this information, if interviewing is not feasible. These stakeholders are affected by the problem or the solution or both. We need buy-in from them. This is a vital step that is often overlooked by people in power. This leads to stockholder ownership of the solution and a more streamlined acceptance and adoption of future innovation. What are the biases of the people who are instrumental in the success of the project? Get the facts and determine what opinions are not factual, and address how these assumptions can be changed.

Is this an old problem? Why is it not working now?

What are the limitations of this problem? This may include funding, human capital, and timing. The goal is to take these constraints and determine if they are indeed the problem, or determine if there is a creative way to work around them. Are there any missing pieces of hardware, software, tools, and so on required to accomplish the task or to interact with the product? Is there a shortage of raw materials such as parts, consumables, information, and so on used to produce the final product?

List even the obvious assumptions and test them for validity. Be a bit skeptical in your attitude. Is this really a fact or is this self-imposed? Is there a way to measure the quality or success of a solution to the problem in place?

### Question 3: How Urgent Is the Solution Needed?

Determining the urgency of the problem means that the process and evaluation of the problem may need to be accelerated. Is it worth solving or will the solution have a domino effect on something along the process that must remain unchanged? Will it go away, if I ignore it? If so, then all other questions are answered.

### Question 4: How Are Your Competitors Dealing With This Issue?

Researching how competitors have solved this issue does not mean one needs to follow suit. Leaders plow their own path. It just gives another perspective. Will the solution be a replication of what exists, but in an improved format, or will it be an original?

### Question 5: Why Did This Happen?

Make a list of about five causes. Review them to make sure they are not symptoms of a bigger problem. Review the current situation rather than what happened before. Here are some examples:

> **Resource causes.** Is the problem caused by resources? Is there a lack of funding? Or are some resources not identified, not used effectively, or applied to an area that is not benefiting from them?

**Human capital causes.** Are there enough people to do the job? Are any of them in need of additional training? Are some people not productive because they lack training, mentoring, or understanding of the process?

**Process causes.** Are the rules, processes, and procedures perceived as roadblocks by the workers? Are the rules, processes, and procedures confusing? Are there specific requirements (i.e., policies, procedures, rules, regulations, and laws) that are causing the problem?

**Environmental causes.** Is the work environment too stressful? Is the environment open to creative thinking and innovation? Are the people in power aware of the problem? Are they supportive of solving it? Identify bias or prejudice. What is the environment that is causing the problem (i.e. culture, location, time, temperature in which it operates)?

**Communication causes.** Is there a special company language filled with terminology, words, and concepts? Are there hidden meanings? Is the communication understood? Is there a lack of communication?

## Fishbone Diagram

Categorization labels often pinpoint who to blame, rather than what to blame. As failures are usually process based, it is more important to determine what caused the problem in the first place. Identifying the cause of the problem gives the information needed to solve the problem.

Observe what is happening today that is causing the problem. Do not look back at what caused the problem in the past. Make a list of the causes and organize and prioritize them.

Many manufacturing, marketing, and service industries like Mazda Motors and Toyota use a derivative of the Ishikawa diagram, which is also known as the fishbone diagram, cause-and-effect diagram, Fishikawa, or herringbone diagram. It is one of the basic visual brainstorming tools in which to sort ideas. The procedure was officially developed by Dr. Kaoru Ishikawa (1915–1989) in 1968 with applications in manufacturing quality control. Each cause is placed in a line that ends in a box identifying a problem, just as the diagram illustrates.[3]

88    IDEA ENGINEERING

### ISHIKAWA's FISHBONE DIAGRAM

*The fishbone diagram, which was invented by Dr. Kaoru Ishikawa, is also called the cause-and-effect diagram, the Fishikawa, or herringbone diagram. This diagram is an example of some factors (i.e. equipment, process, people, materials, environment, management) that may affect the problem.*

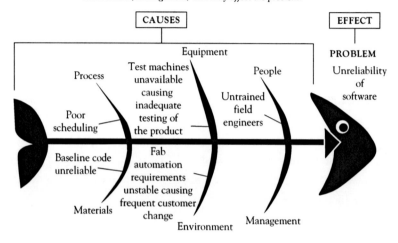

I usually have the team begin by using Post-it notes and jotting down one cause per note and then prioritizing them. Draw a long line on a whiteboard or flipboard, as a symbol of a "fish spine." The "fish head" (a box) is drawn at one end of the line and is symbolic of the problem. The Post-it notes are placed on the fish spine with the most significant problems physically located closer to the fish head or by circling and numbering. If causes are related to each other, connected layers can be added from the initial "ray." There are also programs to do this process on the computer, but in a team setting, a hands-on haptic approach works very efficiently, gives everyone a chance to participate, and opens up the discussion.

## Concept and Mind Mapping

A way to facilitate the relationship between concepts is concept mapping or mind mapping. Concept mapping is a way of organizing ideas graphically, using networks of concepts represented by a web of nodes and lines. A concept map may have several concepts. Beginning with a main idea, add key words and symbols that represent ideas and words that stem from the main idea. It is visually outlining ideas through relationships between knowledge.

Concept mapping, which was developed by Joseph D. Novak and his Cornell University research team in the 1970s, is based on the learning movement of constructionism and the cognitive theory of assimilation, which builds upon prior knowledge.[4]

**INTERNET MARKETING CONCEPT MAP**

Mind mapping is a little more colorful and promotes the free flow of ideas. Unlike a concept map, a mind map has one main concept that can be represented as a tree. This technique is also a way of managing information. Mind mapping was invented and copywritten by Tony Buzan in the United Kingdom to improve readability.[5] Mind mapping is especially helpful for visual and tactile learners. The technique relies on links and associations—nonlinear methods. Key words, phrases, colors, and images are drawn out on paper. The visual and tactile engagement of creating the map acts as a memory aid.

When I was at Purdue University, I used mind mapping as a technique for students to incorporate into their study sessions to improve comprehension. It helps with assimilation and memory, quick reading, and fast review. Al Gore[6] and Bill Gates[7] also use mind mapping on a regular basis to analyze their ideas and order their thoughts.

Fortune 500 companies like Boeing, Oracle, British Petroleum, and Nabisco use mind mapping to communicate and study complex ideas for brainstorming, project planning, and presentations. Boeing used to provide an aircraft engineering training manual for their senior aeronautical engineers. It took a few years for the 100 engineers to learn the content. When the manual was condensed into a 25-foot-long mind map, it took the engineers only a few weeks to learn the content, resulting in an estimated savings of $11 million.[8]

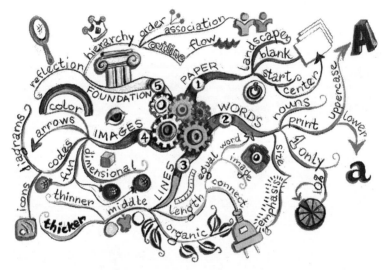

*This is an example of a hand-drawn mind map.*

There are a number of concept and mind-mapping applications available on the Internet. Some are free and some need to be purchased.

## Clear Problem Statement

There is now a better understanding of the problem. Information on the consequences, along with the reasons for the problem have been gathered through observation and research. A sense of urgency has been identified and communicated, and the competitors' situation has been researched. The relationships between the key factors have been determined, and they have been put in order of priority. Now address each key factor as a separate innovation challenge. Because it is not always easy to gather all the facts, assumptions may need to be accepted on faith alone. If they are proven wrong, they should be tossed out right away.

Preparing a written problem statement is the next step. Getting it in a written format formalizes the process. Write the problem statement as a question. Assume that there is not one answer to the problem. "In what ways (verb) (noun) (end result)?" Here is an example: "In what ways do blind and low-sighted individuals understand shape in a haptic environment?"[9]

Identifying the real problem to be solved is the most important step in the innovation process in the business world. This is also true for a graduate student's thesis or dissertation. When graduate students are working on their dissertation or thesis, defining a clear problem statement takes the largest investment of time. If this step is not done correctly, it does not matter how much time is put into the presentation or the prototype. The wrong problem will be solved.

After the statement of the problem is down on paper, look around to find a trusted mentor, coworker, or team to review it. Present the problem statement with a brief of the problem, what has already been tried or suggested, who is affected, and what is the best problem definition. My advice is to engage intelligent people to gather input data. What I mean by this is to interview experts and trusted contacts (which may include a mentor or a trusted outside source) on what they have to say about the problem, causes, and priorities. Describe the problem as objectively as possible, and make any significant changes based on the feedback received.

## A Different Way of Defining a Problem

Amazon's success is partially attributed to their different way of thinking about how to attack problems. The leaders at Amazon view problems as the result of broken processes, not broken people. They view the experts as the frontline workers. "Attack the process. Engage the person," says Emily Thomas, one of the supervisors at the Laveen, Arizona plant (E. Thomas, personal interview, April 17, 2013). Amazon's motto is to address problems immediately, because "Problems are treasures."

The Toyota kata[10] inspired Amazon to do standardized work. It usually takes about two years for companies following the model of the

Toyota kata to get their employees to fully understand, get full buy-in, and change their mindsets. The idea is to standardize a process first so that frontline workers know how to do their job. This does not mean status quo and getting comfortable doing the same repetitive process. It does mean that once it is standardized, there is no variation in the process unless a problem exists; however, the goal is to find the problem along the process to create a better way.[11]

"Adaptive persistence," a phrase coined by Richard T. Pascale, is fitting to describe the learn-by-doing method that Amazon uses to define their problems and improve their process. The official name is PDCA. The process improvement concept was actually introduced to Toyota by Robert Demming. It involves "Kaizen," which is continuous improvement and it is about behavior routines. It translates to meaning that the process is never finished.[12]

Before a shift starts at Amazon, frontline workers review a trifold checklist and ask themselves, "Have you done x, y, or z?" Examples: "Have you reviewed the preshift plan? Is your work area ready to use? Have you discussed the plan with your team?"

At each process stage, there are giant rolling carts with strive cycle charts on whiteboards. Target conditions are noted on the board by the expert frontline workers. The "target condition" means "the challenge" or "the problem." Example: "Create a process that allows the problem solver to respond to a packer about the problem within 10 seconds."

| Name | Process | Target Condition | |
| --- | --- | --- | --- |
| Next Step | What do you expect? | What did you observe? | What did you learn? |
| | | | |
| | | | |
| | | | |

PDCA identifies the challenge, the actual condition today, roadblocks and which ones are addressed now, the next steps, and what has been learned. The focus of the Toyota kata is to truly understand what is going on in the process and to learn about the work system. This understanding and learning is done through observing and studying the system. Applying one change at a time enables workers to see if there is improvement through cause and effect. The current condition and the target condition are measurable. Many companies faced with a problem immediately concentrate on stopping the problem using the first idea that comes to mind, hiding the problem, or jumping into applying several changes at the same time. This often results in a temporary fix or solving the wrong problem, because it was not understood and defined in the beginning. The Toyota kata is not management by objectives (MBO) or return on investments (ROI) metrics. It is a way of working with others by striving for a target condition and respecting the ideas of others.

Communication is key at Amazon. Every week, "5 × 5 × 5" feedback meetings between frontline employees and their process-stage leader occur. When a problem is reported, the process-stage leader observes the process for five minutes, coaches the employee for five minutes, and then observes again for five minutes. Conference calls with other Amazon employees in similar capacities occur on a weekly basis to discuss process improvement. If a stage of the process is changing, the frontline employees are given information as to why this is taking place. This is most important because it helps get employees to understand and support the change. Amazon also moves people around for cross-training purposes, so they can see the big picture and learn how the entire business operates.[13]

At Amazon, as in most businesses, "time is money." The inbound process was standardized first, because customers cannot order an item unless it is on the shelf. So having the shelves stacked is a priority. Originally, the standardized workers packed a variety of sizes of product boxes on their cart to stack. The inconsistent box sizes and locations proved to be a problem for many workers. Quality, speed, and safety were targeted goals. The problem was defined and addressed, and a predictable cycle time resulted. The solution was to put the same number of units on the carts that were similar in size. Now it takes 20 minutes to unload and stow each cart, which has 12 totes per cart. Before, it often took over an hour to unload some carts.[14]

The Toyota kata way of thinking about problems has proven effective. Take a moment and think about it on a personal level. How can this be applied to your life when dealing with uncertainty?

## Food for Thought

The problem has been defined. A deeper understanding has been reached, and much has been learned along the journey. A direction with a focus that is not too narrow has been identified. You have assumed you do not know the answer. The focus is on a need or a service, not an organizational goal. A tangible problem without limiting possibilities has been found. Now you can move forward on your creative thinking journey on the road of innovation.

# CHAPTER 7

# Creative Thinking

*Creative thinking, in terms of idea creativity, is not a mystical talent. It is a skill that can be practiced and nurtured.*
—Edward de Bono

Growing up with a Japanese father puts a fire in your belly when it comes to getting a higher education. I knew I was university bound by the time I was in third grade. According to my father, there was no other choice for me. "They can take everything else away from you, but they can't take away your education!" These words were etched in my brain. My mother, on the other hand, was a musician and an artist and was all about having fun. So using both sides of my brain was something I developed quite naturally.

Early in my career, I was certified to teach art and technology. I have witnessed the changes in the American educational system from kindergarten to graduate school through the past several decades. This chapter is a response to the growing concern that students at all levels are being taught to pass tests and are therefore becoming successful at memorizing factual information and delivering packaged answers. This, in turn, leads to recently hired young engineers and technologists who do not know how to think creatively to come up with ideas leading to innovative solutions. This is the number one concern I have heard from executives and senior engineers and technologists. Today's companies end up paying for this in the long run.

Contrary to popular belief, innovative ideas are not suddenly produced in a frenzy of creative activity. There is a defined step-by-step process before an innovation goes to market: (1) problem definition (which we discussed in depth in the previous chapter), (2) ideation, (3) selecting the best idea, (4) the architecture or building phase, (5) testing and revising, and (6) implementation. In this chapter, we are going to focus on the second phase: ideation, the generation of ideas. Once the problem has been defined, one can use a variety of ways to think creatively to solve that problem.

This chapter covers basic differences between convergent and divergent thinking. It offers warm-up brain exercises and successful techniques of creative thinking such as: (1) brainstorming, (2) lateral thinking, (3) brainwriting, (4) metaphoric thinking, (5) SAMPER, (6) Synectics, and (7) TRIZ. When you apply any of these techniques to your creative thinking sessions, use them as guidelines to adjust as necessary.

The reason this chapter is valuable for engineers and technologists is because in the fields of engineering and technology, one way of thinking tends to dominate. The information in this chapter will teach engineers and technologists how to think like designers, inventors, or entrepreneurs.

## Convergent Versus Divergent Thinking

Students of all ages need to learn how to think. If we want to produce more innovators in this world, we need to tap into both approaches to thinking. Using both sides of your brain is a way of integrating convergent and divergent thinking.

In the fields of engineering and technology, convergent thinking tends to dominate. Convergent thinking is logical and critical and is used when we narrow down decisions. This type of thinking works well with selecting the best idea, after intense sessions of creative thinking have taken place. The next chapter will discuss this narrowing-down process.

Divergent thinking is what engineers and technologists need to develop to be visionaries and innovators. Divergent thinking is the ability to elaborate and to think of novel and diverse ideas. Ideation or idea generation is an example of divergent thinking.

I believe the most important impact a university education has on potential engineers and technologists is to teach them how to think creatively. Taking an original approach in solving problems usually brings about more questions than answers. "Why?" questions are great for inspiration. "But I'm not creative!" is not an excuse. The good news is that creative thinking can be developed and enhanced when it is threaded throughout university disciplines. Idea engineering is about engineers and technologists using creative thinking to form a new worldview leading to innovation.

## Inspiration

Inspiration gets your creative juices flowing. It motivates a person to be creative, to do good, or to feel. Inspiration results from something that happens that causes a person to want to take action.

My husband and I have known a couple—Arlen and Kay Nelson—for many years. They are musicians and play in a band. Arlen's son, Travis, plays the dobro instrument.[1] One day, we got a frantic phone call from Arlen that Travis (age 25) had been in a terrible auto accident and was in a coma. His spinal column was frayed and it looked like it had exploded on the X-ray. Three medical experts told Arlen that Travis would be a vegetable or brain-dead.

Every day Arlen would go to the hospital, sit next to Travis, who was still in a coma, and play his banjo for him. After five weeks, Travis woke up. His right side was paralyzed, but he was alive and alert.

I could not believe it when Travis not only learned to walk and talk again but insisted that he wanted to play in the band again. Travis got on his creative thinking hat and invented a gadget with a strip of Velcro and a guitar pick to wrap around his right thumb. This invention allowed him to play the strings on his dobro even though he had some difficulty moving his right arm. I watched the first performance after the accident. It was impressive.

Travis then decided he wanted to learn how to play the traditional guitar again, but because he had limited control of his right arm and hand, he redesigned a neck clamp guitar capo[2] for his cords using two store-bought capos. He invented a new way of playing notes by holding an 8-inch section of a Volkswagen air cleaner hose in his fist. He positioned the movable capo over the neck of a left-handed guitar and began strumming away with the hose. Pretty inventive. And by the way, Travis also taught himself how to play the fiddle.

Then, Travis invented a gadget with a handle to cradle his large soda to prevent himself from dropping it when he got his car keys out of his pocket. Yes, he is now able to drive again. I witnessed firsthand the invention of useful objects because of the strength of inspiration.

## Inspiring Your Idea Team

Look for inspiration in your idea team. It is easy to observe whether the team is a hot team of innovative people or not.

What does your idea team look like? Are the members of the team engaged, concerned, and empowered? Engaged members are committed to the goal. Concerned members are interested in the project. Empowered members are able to affect change. Energized, productive, enthusiastic, and motivated members lead to action. If this describes your idea team, then you have a potential group of visionaries and innovators!

On the other hand, are the members of the idea team disengaged, frustrated, stuck, and apathetic? Disengaged members are detached from the vision and show no interest in the goal. Frustrated members are ineffective, because they induce feelings of discouragement in the team. Members who feel stuck are trapped, feel restricted, and cannot move forward. Apathetic members are halfhearted and express indifference and boredom. It is doubtful that any fresh new ideas will be created by this team.

When a new idea team has formed, if feedback has not been solicited for a while, or if the team members have been distracted, then it is important to do brain warm-ups. There are numerous ways of exercising one's brain so that it is free to think creatively. This is similar to doing warm-up stretches before running a 10K. One's brain needs the same consideration before the team meets and the ideas begin to fly. Here are some brain warm-up exercises I do to become inspired:

1. **Unwind.** One of the first things I do when I need to be inspired is to go to a quiet place. I go to the Botanical Gardens, or the zoo, or I sit on a bench and listen to water flowing. I refuel.
2. **Have conversations.** For those extroverts out there, start off by networking and having conversations with intelligent and creative people. I do this after I have refueled.
3. **Do not lose your childlike wonder.** Visit a toy store. If creative thinking appears childish or beyond your personal reach, think again. If creative thinking appears to be out of your realm of

expertise, think again. The Wright brothers, who invented the airplane, were bicycle mechanics, not aviation engineers. Kodachrome, which is a brand name for one of the first successful color films used by cinematography and still photography, was invented by two musicians.[3]

Try the Purdue Creativity Test (PCT) and see how many ideas can be constructed from a paper clip.[4] Play children's games or play with their toys. If a survey was taken, one would more than likely find that almost all healthy five-years-olds of normal intelligence test as highly creative, but only a very small percentage of adults do. What happened to us? Did we forget how to play? Do we need to start playing again? Children's games have few rules, and sometimes they even make them up as they go. Play is an important part of innovation. I remember as a small girl, playing with my brothers' Legos and Erector Sets more than they did. As an adult, I have a small set of Legos in my desk drawer.

4. **Participate in routine activities.** Mundane activities allow one to think freely; so walk the dog, walk around the building, clean your desk, but actively think with limited distractions. Make creative thinking a deliberate activity. Just think, it is like killing two birds with one stone (figuratively speaking).
5. **Play cognitive games.** Log on to websites like luminosity.com and do the brain calisthenics designed by neuroscientists to improve one's cognitive skills.

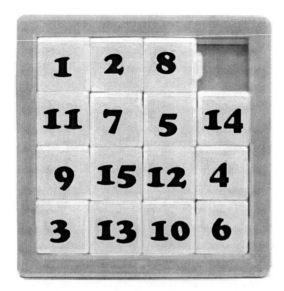

6. **Think in terms of opposites.** Thinking in terms of opposites is a hallmark of creative thinking. The more one learns to be creative, the more tolerant one becomes of opposites. Switching your subject into a contradiction and finding a useful analogy is sometimes called Janusian thinking, named after the two-faced Roman god. When you create two opposite ideas as being true, you are practicing paradoxical thinking (i.e., be sensible, but urge the use of creativity. Build a collaborative team, but accept contradictions). Here is how we do it:

**Problem:** Foundry engineers report that the sandblasting that is used to clean forged metal parts gets into the cavities and is costly and time-consuming to remove.
**Paradox:** Sand particles must be hard enough to clean the parts, but not too hard to be easily removed.
**Essence (useful analogy):** Ice is a particle that is hard but disappears when it melts.
**Novelty:** melting feature
**Equivalent solution:** Make particles out of dry ice to clean the forged metal parts. The particles will later turn into gas and evaporate.[5]

7. **Switch from your dominant brain hemisphere.** Left-brain activity is processed logically and sequentially. It takes pieces, lines them up, and then draws conclusions. Right-brain activity is processed randomly, intuitively, and holistically. It sees the big picture first.[6] When we are learning something new, we prefer our dominant brain hemisphere, because it is familiar to us. The reason it is important to switch ways of thinking is that being too familiar with an idea, a process, a system often makes it so easy that you become overly comfortable in your expertise. This makes one less likely to experiment with different approaches. That is why at research universities I was the gray thinker in a world of black and white. I was there to make engineering and technological minds a little uncomfortable so that they would grow in their creative thinking.
8. **Try concept or mind mapping.** This tactile technique of drawing relationships between concepts is a creative way of taking notes or outlining. It is explained in more detail in the previous chapter. Exercise your brain by outlining a TED talk: "Ideas Worth Spreading" (http://www.ted.com/) video on YouTube.com.[7]

## Idea Conceptualization Phase

Creativity is warranted in the world of the artist, but in the past, it has not been valued outside of what people perceived to be the mystique world of aesthetics, expression, sensitivity, and emotional connection. Today, people recognize that creativity is required for optimal change and progress in the business world.

Conceptualizing ideas is done through the methodology of creative thinking. Thinking of ideas is an art, and creative thinking is a learned skill. It improves with practice. It takes skill and experience to be really successful at participating in an intense thinking session, as well as facilitating a session. For an organization to be innovative, it has to undergo a cultural transformation and provide an avenue for employees to express ideas without rejection or consequences. Always following the old rules, selecting the first idea that pops into one's head, being quiet for fear of failing, or believing there is only one best answer, are sure ways to fail at creative thinking.

## Brainstorming

Alex Faickney Osborn (1888–1966), the originator of brainstorming, was once a journalist in New York, until he got fired for apparently not having the skill and talent to do the work. He ended up in the advertising world and eventually became the executive vice president of BBDO Worldwide, an international advertising agency based out of New York City. After over four decades, Osborn left advertising, because his writing career on the topic of creative thinking was so successful. One of his favorite quotes was "It is easier to tone down a wild idea than to think up a new one."[8]

Brainstorming is the most popular method used to generate ideas. The method was further developed by Professor Sidney Parnes of the Buffalo Institute and is often called the Osborn–Parnes model.

It is so widespread, because it is based on the belief that ideas are generated faster in a group setting. In other words, brainstorming does not occur in a vacuum. Osborn believed in the team activity of using the brain to create a storm of ideas to solve a problem through free association with deferred judgment. Osborn believed that the optimal number of panel members should be 10 to 12, including a facilitator and a recorder. Any more than that would cause people not to participate due to group dynamics or time constraints. The size of the brainstorming group can be slightly smaller if you are limited on available members.

The first objective of brainstorming is to get all the ideas that one can think of on the table. Then modify the ideas or take multiple ideas and combine them into one. Being open to changing directions is important, so that there is no overcommitment to a particular solution when the focus should be on finding the best solution to the problem.

The early stages of ideation, often referred to as the "fuzzy front end," involve decision-making associated with ambiguity.[9] This is one of the most difficult adjustments to thinking that many people have to make. Revolutionary thinking stems from believing that no ideas are bad ideas during the brainstorming session, because brilliant solutions are often discovered in the oddest ways. Creative thinking sees beyond the obvious and uses the imagination to repurpose an idea.

Successful ideation is rigorous and disciplined, as well as creative. Much energy is expended in this process. However, done properly, it is great for cultivating creativity.

Brainstorming sessions are used not only to grow ideas but also to combine and enhance them. What I have found most effective is to come prepared by doing some individual creative sessions before the team sessions. Write down ideas and bring them to the team brainstorming session. Even after the team session, one can do more individual brainstorming to warm up the brain and tweak the ideas. Then bring the tweaked ideas back to the team for further discussion and refinement.

### *Rules of Brainstorming*

Group play has rules. We need these new rules to break the old rules. Here are some specific rules of brainstorming to help the process along:

1. The team leader/facilitator should not be the first one to offer an idea. In fact, the moderator should solicit feedback from the team without offering his or her opinion.
2. Not everyone needs to take a turn. Sometimes, it takes a while for people to warm up to the process and feel safe enough to take risks.
3. Do not have only experts from one area on the team. A diverse mixture of people on a team allows for cross-departmental insight that can help foster ideas.
4. Have both males and females and divergent age groups on the team. Mix it up a bit with people who have different brain hemisphere preferences.
5. Brainstorm on-site. There is no need to go off-site to brainstorm, unless it is more beneficial to be in a new safe environment.
6. All ideas are accepted. Engage absurd ideas, because other ideas will surface and people will be comfortable contributing to the list of ideas.
7. Use markers, Post-it notes, large walls of butcher paper, and so on so that all the ideas are visible to the group. Have a designated notetaker.

I have seen Post-it notes, colored markers, and rolls of newsprint paper on walls and tables used as the foundation of the process. Hundreds of ideas and sketches need to be documented. Often a Show-and-Tell takes place to better understand the concepts. Team members can place Post-it notes on their favorite ones.

*Brainstorming Limitations*

Even though brainstorming is an intense way of problem solving, it is an enjoyable process. There are two obvious limitations to the original brainstorming technique popularized by advertising executive Alex Faickney Osborn: (1) He did not require panel members to do homework before the session and (2) he also offered no exercises to warm up one's brain.

After years of being involved in the brainstorming technique, I have found three other limitations that can be addressed.

1. **Blocking.** The top limitation is that only one person can talk at a time. This is termed blocking. Our human brains have difficulty developing new ideas when we are holding on to our new ideas and waiting to tell others about them. We have an ineffective short-term memory problem. While we are patiently waiting for a chance to

talk, our ideas will be forgotten, edited, or judged. If we get a chance to introduce a new idea, we may get part of the way through the description when someone else jumps in and continues the thought in another direction. As the brainstorming team increases in size, the number of blocked participants increases. This results in fewer ideas compared to when people generate ideas on their own.

2. **Evaluation anxiety.** The second limitation of brainstorming is evaluation anxiety. Some team members are apprehensive about expressing crazy ideas for fear of being judged, especially by an authority figure; so they may be reluctant to participate. Elimination of fear is how creative breakthrough thinking occurs. Ideas are often criticized because they appear childish or not feasible, but this should not occur during a brainstorming session. To deal with perceived criticism, all members of a critique must have an open attitude, because a viable solution can be linked to the most outlandish idea. Remember that constructive discontent is necessary to challenge the status quo. For ways of addressing evaluation anxiety, review Chapter 1 "Getting Your Mind Right."

3. **Personality clashes.** The third limitation is personality clashes. Brainstorming sessions can be controlled by overpowering team members if the facilitator does not manage the session correctly. If team members are more timid, they may only present safe ideas. Stubborn members, who are right fighters (who always have to be correct or have the answers) have difficulty accepting the ideas of others. They also are overprotective about their ideas and take it personally when they are not received well.

Why is it that the arrogant need to always be right? This shuts out creative thoughts. Perhaps the person who needs to be right at all times holds conclusions that originate with an erroneous premise. Try being wrong on purpose to scramble the information, and let it sizzle for a while before casting judgment.

### Examples of Professional Brainstormers

IDEO's brainstorming session has the goal of 100 ideas for the hour session.[10] IDEO's ideation sessions begin with homework. Participants are asked to research the problem to be discussed the night before. They often begin by playing a word game to set the stage for creative thinking.

> - Example #1: **Game:** The Weight of Words
>   Find a fourth word that is connected to three other words.
>   Heart, candy, sixteen = SWEET (sweetheart, candy is sweet, sweet 16)
>   Apple, chart, Jack = \_\_\_\_\_
>   Cow, pie, sun = \_\_\_\_\_
>   Birthday, endings, Lama = \_\_\_\_\_
>   Foot, golf, country = \_\_\_\_\_
> - Example #2: **Game:** 50 Words
>   You are given one random word. Write 50 words about it.
> - Example #3: **Game:** Short Story
>   You are given four random words. Write a short story using all four words: one in the title, one in the plot, one describing the main character, and one describing a twist in the plot.

An ideation facilitator is used to guide the discussion and knows when to change the focus. The focus is on one problem to be solved at a time. Participants come up with wild ideas and plenty of them. When wild ideas are being presented in a brainstorming session, one person speaks at a time, judgment is deferred, and ideas are presented visually and physically as well as orally. The goal is to come up with 100 ideas per hour. Ideas are inspired by other ideas.

Google has eight brainstorming sessions each year with 100 engineers. Six top concepts are pitched for 10 minutes each. The goal is to have one idea per minute inspired by each idea.[11] Some companies have Yes! Meetings in which every idea is greeted with an encouraging "Yes!" to let the ideas flow freely. Apple has paired design meetings (production and brainstorming), and management meetings (pony meetings) each week, and produces more than three times the idea mock-ups than other design departments.[12]

## Lateral Thinking

Lateral thinking is an attitude used for insight and creativity. The human mind has an incredible ability to handle incoming information by establishing a code system that recognizes patterns. Lateral thinking brings insight to the process of breaking through the confines of outdated ideas and brings creativity by generating new patterns.

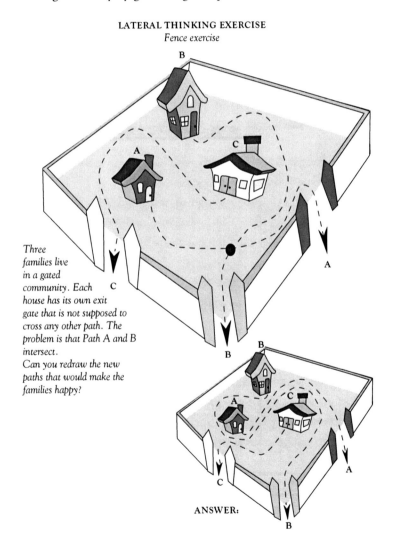

**LATERAL THINKING EXERCISE**
*Fence exercise*

*Three families live in a gated community. Each house has its own exit gate that is not supposed to cross any other path. The problem is that Path A and B intersect.*
*Can you redraw the new paths that would make the families happy?*

ANSWER:

Edward de Bono, the originator of lateral thinking, defines creative thinking using three concepts: (1) One cannot change ideas and perceptions just by trying harder and repeating the same old ways one has been doing things; (2) If one wants to dig a hole in a different place, it cannot be done by continuing to dig in the same place; (3) One can use lateral thinking as a way of jumping across one pattern to another.[13] Einstein was quoted as saying "Insanity is doing the same thing over and over again and expecting different results."

Here are a number of misunderstandings about lateral thinking:

> **Misunderstanding #1:** To be a creative thinker, one has to be an artist.
> Truth: One does NOT have to be an artist to be a creative thinker.
> **Misunderstanding # 2:** One must wait for an epiphany.
> Truth: Do NOT wait for an epiphany. Practice deliberate creative thinking.
> **Misunderstanding #3:** Creative thinking is a trait that one is born having. One is either creative or not.
> Truth: If one is coached and trained, it is possible to go further than if you do nothing about developing your creative thinking skills.
> **Misunderstanding #4:** One has to be a rebel to be successful at creative thinking.
> Truth: One does NOT have to be a rebel to be successful at creative thinking. However, there must be motivation and a deep desire to become creative.
> **Misunderstanding #5:** Creative thinking requires that one only be playful and free in spirit.
> Truth: Being playful and free in spirit is not enough. One needs tools. As adults, we have lived and experienced enough to have developed connected patterns in our minds, whereas children have not. Children can be creative within their innocence. If adults are to be creative like children, adults have to apply techniques to overcome these patterns.

## Lateral Thinking Techniques

If one works consistently on lateral thinking skills, the creative thinking process eventually comes easier and ultimately becomes a habitual way of thinking. De Bono's creative thinking tools include practicing ways to search for alternative ideas and ways to challenge assumptions.

In the process of looking at the many possibilities that lead to the best possible ideas, the search for alternative ideas must be deliberate. Search for as many alternative ideas as possible, regardless of whether they seem reasonable or not. Searching for many possibilities provokes new patterns. This may lead back to the most feasible idea, but time has not been wasted. The process validates the idea that has been ultimately selected, not because it is the only one, but because it was selected from an array of possibilities.

### *Brainstorming Session Using Lateral Thinking*

The ideal brainstorming team size using lateral thinking is six to 15 members, according to De Bono. Larger groups can be broken down into smaller groups and then at the end of the session, their notes can be compared.[14]

Here is how the brainstorming session operates using lateral thinking:

It is necessary to have a team leader (chairperson) and a notetaker. The team leader begins with the problem statement and repeats it during the session, if necessary, to get people back on track. The team leader facilitates the session without controlling it. The team leader's responsibilities include stopping people who are criticizing or evaluating the ideas, stopping people from talking at the same time, and making sure the notetaker has time to write down the ideas. The team leader may introduce different lateral thinking techniques to inspire different ideas. Sometimes the team leader offers ideas. At other times, the team leader asks the notetaker to read the list of ideas aloud, so that ideas are not repeated and also to generate more ideas. Often, the team leader needs to summarize suggested ideas to make sure everyone understands them. The team leader has the responsibility of ending the session at a designated time or if the ideas seem to be dwindling—whichever comes first. The team leader organizes the list of ideas from the brainstorming session and the evaluation session that takes place afterward.

If everyone is talking too fast and he cannot keep up with the list of ideas, the notetaker should ask the chair to slow down the session. One should also consider recording the session. The notetaker is responsible for examining whether an idea is a duplicate of another and should communicate with the team leader if there is a question.

Begin with a 10-minute warm-up session. One can use some of the brain exercises suggested in this chapter.

Research has found that the brainstorming session should ideally be 20 to 30 minutes long. Forty-five minutes is stretching it a little. Stop before every idea is squeezed out of the participants, so that the participants have something to contribute in the follow-up session. The end result is a brainstorming list of ideas, which can be posted online.

The follow-up session can be done using email; however, online collaborative methods, such as Dropbox, Google docs, or SharePoint have more advantages. These web applications allow the idea team to stay in sync by adding, modifying, or deleting content in collaboration with others, as well as exchange large files that can be seamlessly shared. The purpose of the follow-up session is for team members to add more ideas to the brainstorming list and to clarify ideas.

### *Evaluation Session Using Lateral Thinking*

The evaluation session takes place at a later time with either another team or the same team. The evaluation session can be done online or in person. Three lists come out of this session: (1) list #1: useful ideas, (2) list #2: new approaches to the problem, and (3) list #3: ideas that need to be researched further. It is during this session that the usefulness and functionality of the ideas is evaluated. This means that even ridiculous ideas may have a useful purpose.

Select each idea that appears initially useful and put it on list #1. Discuss why it is has value or benefit. Others will add to the comments.

The ideas that are novel or unusual are put on list #2. Examine these ideas and discuss other factors that should be considered. Are there new ways of looking at the ideas on the list?

Look for ideas on the brainstorming list that would be easy to test even if they do not appear to be useful. These ideas that could easily be tested can lead to other ideas. Put them on list #3. Ideas that have actually been tried before can be added to list #3 if it looks like more information needs to be examined to see if this idea is viable. Some ideas on the brainstorming list may need further clarification and more information to be gathered.

The principle of suspended judgment is a clever way of exploring whether the idea is viable or not. It may lead to more ideas. Delaying a decision may result in others offering their ideas. It may determine that the frame of reference needs to be changed. If an idea appears blatantly wrong, switch the discussion and brainstorm how it can be useful in finding the solution.

## Brainwriting

Brainwriting is brainstorming on steroids. A brainstorming technique called "brainwriting" is a successful technique for an open verbal forum. It is often referred to as the "6-3-5 method" (six participants, three ideas each, and five times around the table). It has been proven successful, especially with introverted technologists and engineers.[15]

Everyone sits at a table to address the defined problem at the same time. A piece of paper with the problem statement written at the top is passed out to each of the members. The designated moderator gives each team member three minutes to write down three ideas. As ideas are recorded privately on paper, no one talks to others. Ideas are sketched out or written or both, on the paper. Then when the time is called, the papers are passed to the left, and a new three-minute round begins. Members read what is recorded and add three new ideas. The new ideas can be elaborations of the ones already listed or they can be completely different ideas. Papers are passed around five times and additional drawings and notes are added to the idea sketches. When the time is up or when everyone's brain is fried, the papers get passed one more time to the left. It is at this point that the ideas are read aloud, discussed, and consolidated with the moderator facilitating the process. This is similar to traditional brainstorming. Usually a feasible design will result from this session.

The 6-3-5 method will result in 100 ideas generated in a 20-minute session. No one gets blocked from presenting his or her idea, and everyone gets to contribute equally. Everyone has an opportunity to build upon someone else's idea just like in the traditional brainstorming, only it is done on paper instead of verbally. No ideas are lost, because they are written down. Ideas can be kept anonymous. The challenge and pressure of being given a precise task to satisfy a quota in a definite time allotment is a good way to release creative thinking.

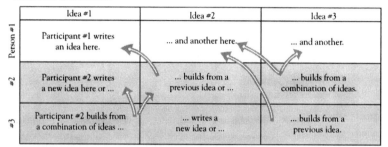

**6-3-5 BRAINWRITING** (in progress)
6 participants + 3 ideas + 5 times around the table

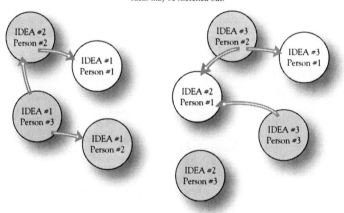

Ideas may be sketched out.

## Metaphoric Thinking

Think like Leonardo da Vinci to change patterns of thinking by connecting the unconnected. Metaphoric thinking is a mental process where qualities of different objects are substituted and compared to each other. In other words, a word or phrase that means one thing is used to describe another, but it cannot be literally applied. An example would be "Time is money." Metaphoric thinking is a way of looking at things with a completely different perspective.

A number of brilliant inventors used metaphoric thinking. Aristotle believed that cultivating a taste for metaphor was a sign of genius. Albert Einstein said that the theory of relativity is "riding on a beam of light holding a mirror in front of your face." The research team trying to understand the concept of superconductivity might say, "Subatomic particles are a dance troupe on the stage." When asked to describe the theory of evolution, Charles Darwin might say, "Evolution is the branches of a tree."

When my daughter Tonya was in elementary school, she just could not manage to get her arithmetic done during class time. The frustrated teacher handed me a sample of her multiplication and I could not help but smile. It looked like this:

She got the right answer, but she certainly had a different way of looking at the solution. A "3" is a butterfly. The multiplication symbol is a kite. The "6" is a cat. The answer "18" is a snowman with an umbrella. Tonya had managed to master metaphoric thinking in the third grade.

On another note, have you ever thought about the theory of the Six Degrees of Separation? This theory, which was introduced by a Hungarian writer Frigyes Karinthy in 1929 through a short story called "Chains," stated that anyone in the world could find a link to another person through a chain of five acquaintances.[16] Pop culture jumped on board with a parlor game based on this concept using the shortest path between a random Hollywood actor and Kevin Bacon. So, apply this to identifying ideas.

The process of metaphoric thinking demonstrates a link between two ideas. A number of years ago, a team of designers from Konica Minolta were in the process of metaphoric thinking and were discussing the best creature that symbolized change. They came up with the chameleon, because it is the master of camouflage and can change its color from one color to another according to its environment. This inspired them to explore colors of light. Organic light-emitting diodes (OLEDS) were discovered when they experimented with the newest display technology and lightweight flexible building materials. They tested plastic and OLEDS, and discovered a new technique of tattooing OLEDS onto the plastic using what we now call an inkjet printer.[17]

## SCAMPER

Another creative brainstorming technique used to improve or invent a product, process, or service is called SCAMPER. It is a mnemonic—a short word to make it easier to memorize.[18] The acronym stands for:

S = Substitute
C = Combine
A = Adapt
M = Magnify or modify
P = Put to other uses
E = Eliminate (or minify)
R = Reverse or rearrange

Alex Osborn came up with the original questions to ask during the creative session. However, Bob Eberle, an educator, took the questions and made them into seven prompts. These prompts formed a mnemonic, making it easier to memorize the technique.

The SCAMPER session begins by taking an existing product, process, or service that needs improvement or perhaps needs to be changed and asking a plethora of questions under each of the seven prompts. Then try to answer them. Here are examples of questions one might ask:

1. **Substitute.** What materials can be substituted to improve the product? Can this product be used as a substitute for something else?
2. **Combine.** What if you combined this product with another? How could you combine resources and talent in the organization to create a new use for this product?

3. **Adapt.** Can this product serve another purpose? What other product is like yours?
4. **Magnify or modify.** What could be changed to modify this product? What could be highlighted to create more value for the customer?
5. **Put to other uses.** How would this product behave if it were moved to another environment? Who else would be interested in using this product?
6. **Eliminate (or minify).** How can this product be simplified? If this product was eliminated, what would take its place?
7. **Reverse or rearrange.** What if the sequence of how this product is assembled was changed? What roles might need to be reversed?

## Synectics

Synectics is a system of creative thinking based on the belief that traditional thought should be challenged. It involves the combination of free-form brainstorming with free use of analogy and metaphor in an informal interchange within an idea team. Synectics is a rational methodology used to solve problems usually in a group setting in which the problem owner (client) actively participates.

Synectics is a psychological approach to creative thinking. The method was founded as an alternative to what some perceived as unstructured chaos that resulted from traditional brainstorming.[19] Synectics sidesteps the thinking process of wildly thinking, deferring judgment, or playing. It jumps right into irrational thinking by using analogical techniques to generate novelty. Because it was set up as "controlled brainstorming," it eliminated the discomfort that often arose for the engineers and technologists when placed in a typical spontaneous brainstorming setting. It also controlled the dominance of some people and formalized the recording of idea generation.

Synectics has three formats: (1) to make the familiar strange, (2) to make the strange familiar, and (3) to create something new. Participants form ideas that detach and then reattach to the problem. Structured side trips or excursions are an important part of Synectics.

William J. J. Gordon and George M. Prince developed this approach in 1960 for business meetings. This process was later developed by Gordon for classroom use. Today, the technique is managed by Synectics, Inc. in Boston. The company trains people in the creative thinking method. It is proprietary and exclusive. The original Synectics process did not have a follow-through mechanism to evaluate, refine, and implement solutions, but that has changed through the years.

## Synectics Procedure

This is a simplified Synectics procedure.

1. Problem briefing
   A clear problem statement is identified by the problem owner (client). Idea team members accept the perception without challenge.
2. Problem analysis
   An extensive analysis of the problem is presented to the idea team by the problem owner. This also includes what has been tried before. Idea team members ask triggering mechanisms (questions).
3. Springboarding
   The client asks for a specific problem to be discussed in what is termed "springboards" or goal wishing. These alternative perceptions of the problem are expressed without challenge. Absurd ideas are encouraged. Einstein said that "unless at first an idea is absurd, there is no hope for it."

   The team playfully comes up with over a dozen "I wish" or "What if" statements that are read out loud with no criticism. Build on the ideas of others. These may be foolishly impractical and absurd. "I wish you could hire someone to take care of the mundane tasks" may lead to "What if a robot could be flown in every time we need extra help?" These unorthodox wishes are not intended to be final solutions, but rather ways of springboarding into the actual solution. One or two wishes rise to the top and are sometimes combined. By not focusing on the problem, but rather the quickest wishes that come to mind, the one wish becomes a potential solution.

4. Problem restated

   The problem owner and idea team participants restate the problem in terms of one single concrete target.

5. Shred the known stage

   Known ideas and initial ideas that come to mind are collected and recorded.

6. Analogy stage

   Analogies move ideas away from the original problem statement and make a forced fit to the restated problem. Analogies are used for inspirations for new ideas. Team members identify relevant analogies. They use analogy and abstraction to remove the problem.

   - Create direct analogies. How are A and B the same?
   - Describe personal analogies.
     Select a direct analogy, become the object chosen, and describe what it feels like to become that object. What would it feel like to be a tractor?
   - Identify compressed conflicts.
     Pair words from personal analogies that appear to be conflicting. Discuss why they are conflicting. Then choose one compressed conflict pair. Example: How is a tractor both powerful and serene?
   - Create a new direct analogy.
     Select a different direct analogy by selecting something that is described by the paired words. Select the best analogy.

7. Excursion time

   The problem owner discusses the idea he or she wants to see evolve. The participants may visit a robotics company or an airport—any place that helps the idea team make connections.

8. Pointwise response stage

   The ideas remaining are assessed for new possibilities. They are evaluated by the team using a method of itemized response (pointwise response). Benefits and drawbacks of each idea are discussed. Try to turn drawbacks into positives.

118    IDEA ENGINEERING

> 9. Revisiting the original problem
> Go back to the original problem. Produce a description or product that utilizes ideas generated in the entire process. Lots of interactivity takes place, and ideas begin to narrow and are developed into concepts.

The idea session is ended by the problem owner when he or she commits to implement the idea he or she selects. The decisions are made by the problem owner alone. He or she agrees to validate his or her understanding of the idea before evaluating it and give the idea the benefit of the doubt. If the idea is not yet possible, because of a shortcoming, the area that needs improvement is identified (such as cost), and the focus of the idea team is on producing more ideas to counter the shortcoming. This results in either a new idea or an alternative to be implemented.

Creating a paradigm shift using Synectics takes more time and is more demanding than lateral thinking or traditional brainstorming. It also takes the commitment of the problem owner.

## TRIZ

Whereas creative thinking methodologies outside of the box, such as brainstorming, brainwriting, Synectics, and lateral thinking, are based on the large flow of ideas leading to a quality idea, TRIZ is a closed-box approach to innovation. TRIZ, which is a Russian acronym, is a more structured and more complicated approach based on the scientific study of the patterns created by problems and solutions. In spite of this, these patterns have resulted in over three million patents codified within TRIZ. The TIPS (Theory of Inventive Problem Solving) approach is another name for TRIZ. G. S. Altshuller and his Russian colleagues developed a refined TRIZ method of inventive problem solving between 1946 and 1985.[20]

The TRIZ technique is based upon the understanding that someone somewhere else in industry or the sciences has probably solved a problem close to this specific problem. To find out what they did, analyze the use of the TRIZ databases for the solution, and then adapt it to solve our specific problem.[21]

Here is an example: The Food, Conservation, and Energy Act of 2008 mandated researching the uses of animal manure.[22] One such use was as raw material for energy production. Dairy farmers were financially affected by the increased cost of energy, so they wanted to figure out a way to dry cow manure to be used as fertilizer without using heat. The TRIZ process led to researching several drying methods such as the one used for the concentration of fruit juice. This solved the problem.

A more simplified technical innovation strategy called SIT (Systematic Inventive Thinking) is a thinking methodology for innovation derived from TRIZ. SIT was developed in Israel in the mid-1990s. It is based on examining the common logical patterns found in inventive solutions, and it has a more practical application than TRIZ. It is easier to learn and remember and can be applied to more disciplines other than engineering.

Embracing limitations can spur creativity. View Phil Hansen's "Embrace the Shake: The Power of Limitations" on TED.[23] The thinking process is a Closed World principle, which runs counter to thinking outside the box. It dictates the use of resources that already exist in the product or its surroundings. It is based on forcing problem solving within limited resources, and at times facing the path of most resistance.

## Food for Thought

Creativity and imagination make us great thinkers and problem solvers. Creative thinking can be taught and learned. It requires organized steps and a deliberate thinking process. There is not one right way of creative thinking. It is the process that is important. Different schools of creative thinking borrow from each other. That is why there are overlaps.

Creative thinking cannot be divided into separate sectioned compartments. It needs to be threaded throughout the organization. It needs to become part of the culture. A culture of risks that accepts insane ideas and an occasional failure is a ripe environment for creative thinking and innovation.

Phil McKinney, Vice President and Chief Operating Officer at HP, fosters innovative ideas by being receptive to anyone's ideas in the company who has a sponsor and an advocate. Innovation does not just happen in R&D at HP. It is a companywide process. HP approaches resistance to the idea by having systematic goals and milestones during the incubation period of the idea.

Jim Collins and Jerry I. Porras, authors of the book *Built to Last*, stress the importance of experimentation and "purposeful accidents" in the creative thinking process. They believe that visionary organizations are successful, because they place a higher priority on strategic planning than their competitors. 3M suggests that successful ideation revolves around the persistence of new and unique ideas and implementation of many inexpensive and small experiments. The experimental part of the incubation phase focuses on the goal of solving problems, not implementing a specific solution. *Keep shifting solution paths until one works* means that for innovation to occur, one must not make the mistake of being overly committed to one solution.

Think about this: Everything one learns in life can be applied to the creative process. Creative thinking is what makes us human. We all have the capability to be creative. We just have to learn to think differently. All of these thinking techniques can be used alone or combined. But remember, creative thinking is more than just generating ideas. If ideas do not lead to action, the ideas will hide quietly in our mind or exist as drawings in an elegant and versatile sketchbook with 70-pound chemical-free recycled paper.

# CHAPTER 8

# Selecting the Best Idea

*Creativity without a goal—tacit or explicit—can become art, but will never be innovation.*

—Juan Cano-Arribi

The beginning of the innovation cycle is the lightbulb moment, and ideas are plentiful. The front end of the pipeline allows opportunities to explore and play. Quality is not an issue, and the time investment is low. The mantra is "fail early and fail often." If an idea does not pan out, learn from it. It is better to learn from mistakes than to insist on being right and cost the company time and resources. Once all the facts have been gathered, the problem has been identified, ideas have been narrowed down to a few, and the selection criteria has been agreed upon, the next move is to **converge** by selecting one idea that will be the most successful at solving the well-defined problem.

In this chapter, idea selection criteria and tools are introduced to explain how to select the best idea—the potential solution to the problem. This process can be referred to as transitioning from divergent thinking to convergent thinking or from inspiration to perspiration. This is not the time for exploring more options.

Having analytical tools for idea selection will feel comfortable for the engineer and technologist. The value of the knowledge in this chapter is learning to work with others to gain consensus. This is a time for closure and making a decision on which idea will be developed.

## Idea Selection Preliminary Work

Before the selection process begins, the following preliminary work needs to be communicated to the people making the decision: (1) the defined problem, (2) the two to three potential solutions, (3) the user profile, and (4) the selection criteria. The decision-makers may be the idea team, another internal group, or an external group.

### The Problem and Ideas

Begin by restating the problem that needs to be solved. It is beneficial to have the problem statement written out so that the decision-makers have it within view. Then list the two or three possible ideas that should be further developed and implemented with a brief description of each.

### User Profile

Before screening for the best solution for the problem, put together a user profile using the information previously gathered about the people who will use the product, process, or service. Have this description available for the decision-makers. By closely listening to the interview responses from the user/client, so much can be learned about what they need and want in a product, process, or service. Just a reminder: If not enough time is spent profiling the user, more time will be spent in the redesigning phase, resulting in higher costs for the company.

### Selection Criteria

We now know what the problem is, what the alternative solutions are, and who will be directly involved in using the innovation. Now how the idea will be judged needs to be determined. Creativity is not just found in the process of ideation (idea generation); it is found in the process of defining criteria for choosing the idea that moves forward. This process helps define our value system and also what the company deems most significant.

The following is a list of criteria used to determine the best solution for the problem. Select a maximum of five criteria from this list or add your own:

- **Appeal**
  Do the innovators have a deep understanding of the human element? What makes things appealing and attractive?

- **Feasibility**
  Does the system or device obey the laws of nature? Is it doable?
- **Viability**
  Does it generate revenue that is greater than the cost of producing it so that it will be sustainable in the marketplace? Can this product be mass-produced? What is the cost/benefit ratio? Is the overhead low? Does the idea have low switching costs and ease in adoption?
- **Usefulness**
  Does the idea satisfy an unmet need? Does it remove an important roadblock in the process or service? Does it solve the problem effectively? Is it practical, reliable, and consistent? Is it an appropriate response to the problem?
- **Aesthetics**
  Is the physical appearance of the product an important marketing aspect? Does the visual aspect of the product make it easier to use?
- **Ergonomics**
  Does the design of this product make it efficient and safe to use?
- **Accessibility**
  Is it important to make the innovation accessible to different genders, body sizes, ages, and disabilities? For instance, if a website or interactive media is designed, is it accessible to employees and the public with disabilities?[1]
- **Stability**
  Is the strength to stand or endure an important criterion for this product?
- **Suitability**
  Does the idea match company goals? Is it a good fit?
- **Resources**
  Are there resources available to get the project developed, tested, and implemented? Does it meet the constraints? Does the idea have company buy-in?
- **Social responsibility**
  What is the social impact of this idea? Does it benefit the economy, culture, society, or environment? Is the idea ethical? Will this idea

lead to unintended consequences and an ethical dilemma? Does it have the potential to do harm?

- **Benefit to user/client**
  Is it affordable? Is it easy to use? Is it acceptable to the user/client? Does it improve the quality of life or work? Does it cause new problems? Are the trade-offs (negative side effects) acceptable? Is the brand's reputation (company image) improved by this idea? Does the idea enhance the competitive advantage? Is the morale of the workers increased?

- **Project leader**
  Is there a strong project leader who can see this project through development, testing, and implementation?

- **Market status**
  Is the idea a new one in an existing market? It is easier to introduce a new idea in an existing market than it is to introduce a new idea in a new market.

- **Customer status**
  Is the idea going to be offered to an existing customer? Introducing new ideas to existing customers is easier than introducing new ideas to new customers.

- **Flexibility**
  Is the idea scalable? Can it be adapted? Can the idea be extended to the customer experience, the business model, the service, or the process?

- **Intellectual property**
  Do patents protect the intellectual property of the design? If so, is there a strong opportunity for intellectual property to act as a barrier for future competitors to copy the idea?

## Screening Tools for Selecting the Best Idea

There are a number of screening tools that can be used to select the top solution to a problem. Some of the techniques are group activities, some are individual activities, and some are a combination. Some require a time commitment and some have a fast turnaround.

All of the following idea selection tools are based on logical and linear thinking: (1) Idea Criteria Tool, (2) Delphi Tool, (3) Priority Tool, (4) Voting Tool, (5) AFV Tool, and (6) Pros & Cons Tool. These objective techniques work if the team is at a standstill, if one idea does not stand out to the team because there are several great ideas, or if there is a disagreement within the team as to which idea to move forward.

## *Idea Criteria Tool (Group Consensus)*

The Idea Criteria Tool is a screening process to weigh the alternatives based on how each idea meets the criteria for which it is being judged. The group agrees on three to five criteria, and the weight (importance) of each criteria. This idea selection technique can be done as a group consensus. It does take time commitment to allow discussion. Discussing the importance of each criteria should be open and inclusive, so that everyone is on board with the weight of each. This process helps with understanding any compromise that results as well as the reasons for the decision.

**IDEA CRITERIA TOOL**

(STEP 1) IDEA #_____      Weight and Score: 1 to 5 (5 is highest)

| (STEP 2) CRITERIA | (STEP 3) WEIGHT | (STEP 4) SCORE | Criteria Score Formula | (STEP 5) Subtotals CRITERIA SCORE |
|---|---|---|---|---|
| A |   |   | A weight × A score |   |
| B |   |   | B weight × B score |   |
| C |   |   | C weight × C score |   |
| D |   |   | D weight × D score |   |
| E |   |   | E weight × E score |   |

A little algebra used here: (STEP 6) TOTAL CRITERIA SCORE =

(STEP 7)
$$\frac{\text{TOTAL CRITERIA SCORE}}{25 \text{ (which is 5 criteria} \times 5 \text{ pts.)}} \times \frac{X}{100} = \_\_\_\_ \% \text{ for Idea \#}\_\_\_\_$$
(STEP 8)

Instructions for the Idea Criteria Tool (refer to the Idea Criteria Tool figure):

**Step 1:** Select the idea to be screened.

**Step 2:** Select three to five criteria.

**Step 3:** As a group, discuss and give a point value to the weight (importance) of each criteria, using the following guideline:

5 = very important
4 = somewhat important
3 = neutral
2 = somewhat insignificant
1 = insignificant

The notetaker records the number that the majority agreed upon. A whiteboard or flipchart can be used.

**Step 4:** Add a score on how each idea supports each criteria, with 5 being the highest rating.

5 = very supportive
4 = somewhat supportive
3 = neutral
2 = somewhat unhelpful
1 = unhelpful

**Step 5:** Multiply the weight of the criteria times each score to get a criteria score subtotal.

**Step 6:** Add all the subtotals for each criteria, and the result will be a sum of the total criteria score.

**Step 7:** Now algebra is used. Divide the total criteria score the group decided upon by the highest possible total score. The figure illustrates the use of five criteria, but if three criteria are listed, multiply 3 by 5 (the highest score); 15 would then be the highest possible total score. To get the percentage, multiply the fraction by x/100.

**Step 8:** Compare the percentages of each idea. The idea with the highest score should move to the development phase.

A team discussion should follow, addressing how to overcome any concerns, who will be affected by the solution, when the idea will be implemented, and so on.

### *Idea Criteria Tool (Individual Ratings)*

Use the same process as the group consensus. The difference is that there is no formal group discussion, and each person fills out the matrix individually and sends the form to a designated point person. The scores are totaled and then divided by the number of team members to get an average percentage for each idea alternative.

### *Delphi Tool*

The Delphi study is a tool that can be used to select the best idea. It is a consensus-building forecasting technique.[2] It uses an external panel of experts to get input. The time investment is higher, because it takes time to recruit participants (usually around 20 people), co-ordinate and communicate with an outside group, and collate the feedback in several rounds to gain consensus. This method can be implemented virtually. A person is assigned to gather the feedback and crunch the statistics.

The Delphi method of gaining consensus begins in the form of a questionnaire with anonymous responses, which also includes experts adding reasons for the decisions. The experts may be asked about what criteria they believe to be important in judging each idea, which idea is the best, and so on. The participants score their responses on a given scale (1 to 5). The facilitator gathers the data. Then the compilation of the data is sent to the experts in the form of the group scores, the number of people scored in each column, a summary of the forecasts (decisions), and the reasons for the decisions. The experts are encouraged to read the feedback from the group and perhaps re-evaluate their position and change their mind, if necessary, to arrive at an expert consensus.[3]

| DELPHI TOOL | | | | | |
|---|---|---|---|---|---|
| Round 2 Consensus: *You will be receiving a Round 3 Survey with a shortened list.* | | | | | |
| These are the suggested Engineering Graphics basic skill sets required for the full-time interns. | Regional Votes | | | | |
|  | A | B | C | D | E |
| Freehand sketching | 24 | 12 | 15 | 45 | 60 |
| Computer-aided design (CAD) | 65 | 75 | 80 | 25 | 45 |
| Fundamentals of engineering graphics | 5 | 50 | 45 | 23 | 35 |
| Drafting techniques | 67 | 23 | 75 | 62 | 45 |
| Lettering | 5 | 7 | 24 | 8 | 5 |
| Orthographic projection | 50 | 43 | 27 | 34 | 35 |
| Sectional views | 56 | 23 | 30 | 25 | 35 |
| Pictorial drawings | 34 | 30 | 27 | 25 | 38 |
| Dimensioning | 20 | 45 | 12 | 43 | 28 |
| Industry practices | 26 | 15 | 35 | 5 | 3 |
| *Add any comments here.* | | | | | |

*This is an example of a summary of Round 2 forecasts.*

## Priority Tool

The Priority Tool originates from computerized bubble sorting. In this idea selection method, Post-it notes or index cards are used. This method takes a moderate time commitment and can be used in a group setting or individually.

The process begins by identifying one to five of the selection criteria that will be used to prioritize the best solution and rank the criteria in the order of importance with the highest number being most important. Label each criteria item with a number on the Post-it notes or index cards. Place them vertically on the left side (refer to the Priority Tool figure).

Label each idea with an alphabetic letter in no particular order. Place the lettered cards in a horizontal row (i.e., ABC) for each criteria.

Begin on the left with the first criteria item. Take two or three ideas and rank them according to how they successfully support each criteria. The rankings are 5 points for the top solution, 3 points for the mid solution, and 1 point for the bottom solution. Multiply the criteria points times the idea-ranking points. Add the scores for each idea. The highest score is the selected idea to develop. For example:

| | PRIORITY TOOL EXAMPLE *Refer to instructions in text.* | | |
|---|---|---|---|
| *(5 pts)* TOP CRITERIA | *(5 pts)* TOP SOLUTION | *(3 pts)* MID SOLUTION | *(1 pt)* BOTTOM SOLUTION |
| 5 | A  $5 \times 5 = 25$ | B  $5 \times 3 = 15$ | C  $5 \times 1 = 5$ |
| 4 | B  $4 \times 5 = 20$ | A  $4 \times 3 = 12$ | C  $4 \times 1 = 4$ |
| 3 | C  $3 \times 5 = 15$ | B  $3 \times 3 = 9$ | A  $3 \times 1 = 3$ |
| 2 | A  $2 \times 5 = 10$ | B  $2 \times 3 = 6$ | C  $2 \times 1 = 2$ |
| 1 | A  $1 \times 5 = 5$ | C  $1 \times 3 = 3$ | B  $1 \times 1 = 1$ |
| *(1 pt)* LOWEST CRITERIA | Adding the overall scores: A: 25+12+3+10+5= 55 (Solution A is the top choice) B: 15+20+9+6+1 = 51 (Solution B is the mid choice) C: 5+4+15+2+3 = 29 (Solution C is the last choice) | | |

## *Voting Tool*

The Voting Tool is an easy group idea selection activity with fast turnaround using a democratic process. After open discussion about the potential solutions, the voting process begins. There are several ways of implementing this voting process: (1) sticky dots, (2) a tally, (3) anonymous voting, or (4) online voting. If sticky dots are used, they can be different colors to represent different values or different company roles. If a tally approach is implemented, flipcharts or whiteboards can be marked with numbers, ticks, checks, or crosses. If anonymity is important so that members are not persuaded by peer pressure, blind voting on paper works. Sometimes it is more convenient to set up online voting (i.e., SurveyMonkey, myDirectVote, etc.), especially if an incubation period is needed before the final vote. The benefit in voting online is that data is collected and can be saved. Some online applications (such as tricider.com) allow idea teams to collect ideas, add pros and cons, and vote online.

*AFV Tool*

The three-part AFV Tool is a logical selection process that focuses on the appeal, feasibility, and viability of the ideas. The process can be implemented as either a group or an individual activity and has an average time commitment. For each possible solution, make a list. **A list:** What makes a product, service, or process **appealing**, desirable, or attractive to people? Why would people choose this innovation? **F list:** Is the idea something that is **feasible**? Can the idea be realized? Can it be put into practice within the allotted time with the allotted human capital? Does the idea have buy-in from decision-makers? **V list:** Is the idea economically **viable**? Is it sensible in cost to develop and implement? Does the idea have the potential for economic growth? Select the idea with the most potential.

*Pros & Cons Tool*

The Pros & Cons Tool is a method of understanding the benefits and challenges of an idea. It is a way of working through any misunderstandings of the benefits and drawbacks of the ideas. This approach is usually done in a group setting and has average turnaround. Each possible solution to the problem is listed at the top. A line is drawn down the middle with the columns "pros" and "cons." All of the positive aspects of the idea that add value are listed in the pros column. All of the reasons the idea is a negative choice are listed in the cons column. Discussion continues on the value of each pro and con and which idea has the most benefits. Sometimes, voting is used to make a decision.

## Food for Thought

Unless an idea stands out unanimously, convergent techniques and tools are used to thin down the selection for consensus. Then the best idea can move forward to the architectural phase of innovation.

If at the end of the day there is no best idea, then consider that the idea selection process could have been defective. The process may have to be done again using another technique. Doing nothing is also a valid decision. Because the language of business is money, the economics of the idea will need to be addressed, but this is outside of the scope of this book.

Part of what makes innovation so intriguing is that the solution to the problem may be only the beginning of many more improvements. Just because something works does not mean it cannot work better. Often noteworthy changes happen quite by accident, when unrelated ideas are associated, and new relationships between ideas are discovered. Even if a product is improved incrementally in cost, quality, and time to market, it is still a significant step of innovation.

# CHAPTER 9

# Designing and Testing an Idea

*Genius is one percent inspiration and 99 percent perspiration.*
—Thomas Edison

The inspiration—the idea—is only the beginning. Now the real work begins. The reason so many brilliant people fail is because they get to the one percent inspiration and are not willing to put in the hard work to bring the idea to fruition. Work ethic and downright perspiration and hard work are what separate those that do from those that do not. Without applying the idea to something useful, it is just a concept floating in the sky.

This chapter briefly reviews project management and the basics of the designing and testing of the idea before implementation. In the design phase of the innovation cycle, plans and decisions are made about the new or improved product, process, or service through visually demonstrating how it will work. The testing phase is a process of gathering information to determine if the product, process, or service does what it is supposed to do. Examples of renderings, mock-ups, and prototypes will be shown, followed by a brief review of testing methods.

The grand takeaway for engineers and technologists is to understand that design is a distinct discipline that involves special skills, that testing gives the design validity, and that design and testing are essential parts of the innovation cycle.

## Doing the Hard Work

This is a perfect segue for me to interject a little anecdote about perseverance. My husband Carl has the reputation of being highly focused and not being able to do things in moderation. This is why he was so successful as a football and track coach. The one hobby Carl loves is to play his clawhammer banjo. Music is a connection to his soul and gives him hours of pleasure. Music inspires him.

Shortly after we arrived at Purdue University, he started having problems with his right hand. It started cramping up on him. It probably did not help that I would try to make him smile by calling him "the claw." LOL. On a serious note, because that is the hand that picks the banjo, this had the potential of being a crisis. He went to a variety of doctors and drove to Chicago to see a specialist at Northwestern University. Carl was diagnosed with dystonia, a neurological movement disorder that causes involuntary contractions of the muscles. There is no cure, only the treatment of Botox at $1,000 a shot.

At first when I saw him packing up his collection of banjos and selling them, I was concerned that he was so disheartened that he was giving up. Much to my surprise, several left-handed banjos arrived at our doorstep. Carl learned to play the banjo from scratch, switching hands. That was amazing to me. All I could think of was how many new brain and nerve connections he was growing in this process.

Today he enjoys the left-handed banjo jamming regularly with a group of old-time musicians in Phoenix. His music has not lost a beat. That is called perspiration. Carl calls it "just plain old sweat."

## Project Management

With perseverance comes strategy. Before deciding how a project is going to be designed, one needs to determine a strategic plan with a list of all the tasks that need to be accomplished, the areas of responsibility, and a timeline for the design team. The project manager's responsibility is to organize these tasks and see that they are carried out. Project management software is available to plan, manage, and organize this process to keep the design team on task in meeting the deadlines. Some are web based, proprietary, or hosted on the premises. Some project management software is set up for internal use only, and others are collaborative and link to global partners. Some project management software is geared

## THE ARCHITECTURE OF AN IDEA

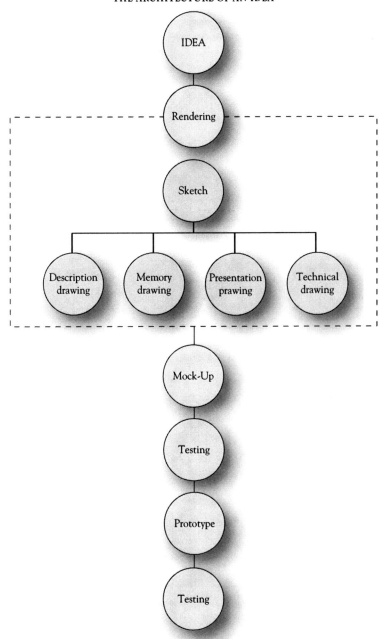

for specific professions, such as a time-location software for engineering infrastructure (i.e., highways, water engineering, transmission line building). Some track the workflow of projects as well as resource allocation of time, money, human capital, and equipment.

## Renderings

The rendering stage of the building process involves hand sketches and drawings produced by detail designers. Bill Buxton's book *Sketching User Experiences: Getting the design right and the right design* is a great resource for renderings as well as understanding the role of design in your own company.[1] This rendering stage can be applied to all types of engineering and technology innovation. Any of the following types of rendering can be used, depending on the end product, device, system, or process.

### Sketch

The first type of rendering is a sketch, which is produced as a part of the thinking process during the early ideation (idea generation) process. Sketches are significant **conversations in the communication of ideas**. Hand-drawn sketches have a number of unique benefits. They relate ideas to team members, users, or clients more efficiently because you can talk about your idea as you sketch. Hand sketching separates concepts from details, is a way to quickly transfer a thought, and captures an idea for future reference. The sketching can be done on a napkin, paper, or a digital sketching program on a portable pad. Sketches can be reproduced in quantity at minimal expense. Sketches are quick, rough, fluid, explorative, and disposable.

A big mistake is rushing to a computer too quickly to produce detailed images before refining ideas by sketching. This results in missed opportunities. I have always believed that the limitation of sketching for initial concepts is your imagination, and the limitation of using the computer for basic ideas is tied to your technological ability. In other words, sketch out concepts before producing detailed work on your computer.

DESIGNING AND TESTING AN IDEA    137

Here are some examples of sketching:

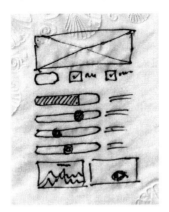

*This sketch was done quickly on a napkin to convey an interactive idea.*

### Memory Drawing

To depict and record a more detailed visual concept, a memory drawing would be produced. Here is an example of a memory drawing:

*This sketch of the character El Dios Verde by Michael Hoerter. Mentor: Dr. La Verne Abe Harris, Purdue University.*

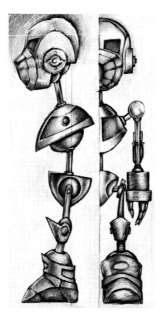

The hand sketch of a robot character was drawn by Professor Ruy Hassan, Purdue University.

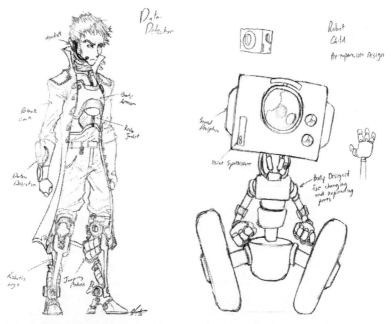

The hand sketches of robot characters and the Data Detector for a proposed interactive game about computer programming were drawn by Michael Hoerter, Purdue University.

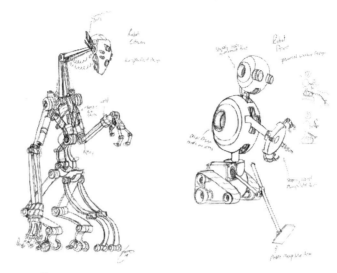

*(Continued)*

### Presentation Drawing

The presentation drawing is produced for those people who need to see a more refined drawing to understand the concept. This is usually reserved for the client, customer, or another stakeholder for presentation and buy-in purposes.

*The 2D presentation character of El Dios Verde was drawn by Michael Hoerter. Mentor: Dr. La Verne Abe Harris, Purdue University.*

*The 2D presentation characters of Samantha and Sabrina were produced by Hana AlJaberi, using a vector object-oriented software. The characters are in the interactive game called "Samantha's Song," a game for children with spina bifida. Mentor: Dr. La Verne Abe Harris, Purdue University.*

*The presentation drawing of the Kart was produced on the computer using 3D software. By Nick Rohe and Joe Gerace; Mentor: Dr. Nate Hartman, Purdue University.*

## Technical Drawing

A technical drawing is produced as an accurate drafting/blueprint communication for builders.

DESIGNING AND TESTING AN IDEA    141

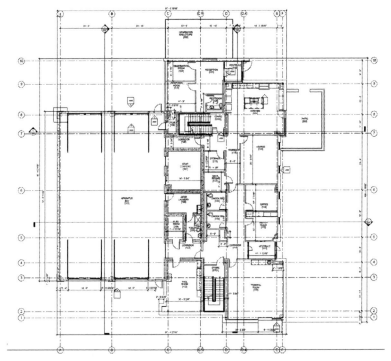

*Kokomo Fire Station #2, first floor. Designed by Hillary Kaub and Nathan Keene. Mentor: Professor Clark Cory, Purdue University.*

## Description Drawing

A description drawing includes visual information on how the innovation works or how it is constructed. It often includes exploded drawings or cutaways. Another example of a descriptive drawing is an airline safety card.

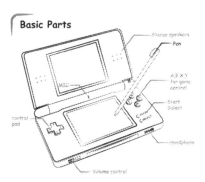

*This is a section of an informational graphic on the Nintendo DS Lite by Yingzi Zhang. Mentor: Dr. Estaban Garcia, Purdue University.*

142    IDEA ENGINEERING

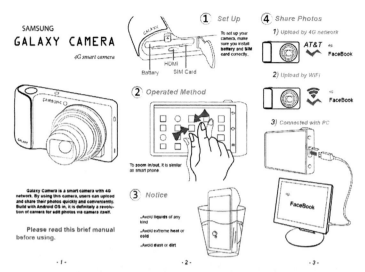

This is an informational graphic with description drawings of a camera drawn by YuFan Song. Mentor: Dr. Estaban Garcia, Purdue University.

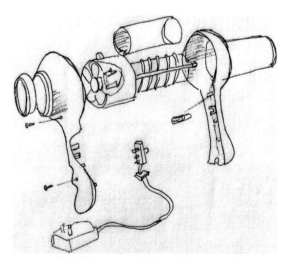

This is a quick descriptive sketch of a hair dryer by La Verne Abe Harris. It was done in about a minute.

## Mock-ups

After concepts have been drawn out, the idea is produced in more detail as a "mock-up," which is the first approval process. A mock-up is the concrete idea in its early stages. It is a rough idea. It is a graphical representation of the idea used to gather data and test for design or feasibility or both from the user or customer. It demonstrates how the user will interact with the innovation.

Mock-ups are inexpensive and easily revised. The purpose is to obtain feedback about the design of the product, process, or service in the early stages of development. Then it is analyzed and corrected. Problem areas are identified, the designers go back to the drawing board, and it is redesigned.

Some mock-ups are called "low-fidelity" or "paper mock-ups." The main purpose of a low-fidelity mock-up is to communicate function. If the product is three-dimensional (3D), a nonworking model could be made using foam, wood, cardboard, or paper, just to give a general idea of the form and size. Mock-ups may be produced using whiteboards, tubing, duct tape, hot glue guns, Legos, foam core, cardboard, a computer animation, photography, and so on.

Here are examples of low-fidelity mock-ups:

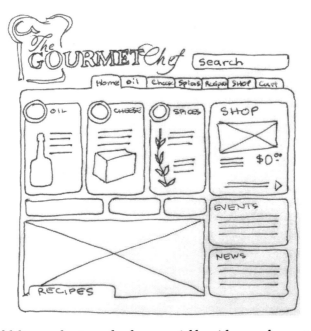

*A low-fidelity mock-up can be drawn quickly with a marker on paper. Interactive paper mock-ups are used to test users' responses before investing money in building the interactive product. Drawn by La Verne Abe Harris.*

User experience (UX) design² is a conversation between the designer and the user in the form of a low-fidelity product, which may be made of paper, cardboard, or other inexpensive materials.

A low-fidelity mock-up, such as a paper flipbook, is way of reviewing the sequence of animation frames. Sometimes, screen shots of hand-sketched storyboard frames are placed in a PowerPoint presentation or produced as an animated.gif, to get a feel of what the animation will look like.³ An animated.gif image can also be produced to demonstrate interactivity or a process change. A low-fidelity mock-up using motion video with cardboard props is an inexpensive, fast, and disposable way of showing role-playing and the change in a process or service.

DESIGNING AND TESTING AN IDEA    145

The player plays the role of the Data Detector and examines the software for bugs. A software fundamentals Q&A session takes place.

When the Data Detector gives an incorrect answer, a bug is released.

The Data Detector aims directly at the bug and shoots by selecting one of the possible correct answers.

One major bug is eliminated, but the minor ones remain. The Data Detector continues the process.

*In interactive design, a storyboard mock-up describes the game interactivity and story line. This storyboard is for a potential video game to learn the fundamentals of computer programming. Illustrations by Michael Hoerter, Purdue University.*

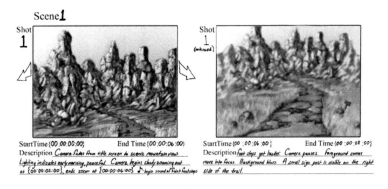

*In animation, mock-ups are usually rough hand-drawn storyboards that are like comic books illustrating a sequence of events about a story. Sometimes they are high fidelity and done in color for the client presentation. They include arrows, the properties of speed and direction, and captions to help understand the transitions. Rough storyboards are also used for more than animation films; they can be used to show interface design steps, to illustrate how a product gets used, how a product works, or how the service or process has changed. The panels are drawn by Professor Ray Hassan, Purdue University.*

*These are examples of illustrations of high-fidelity storyboard panels for an animated film "The Adventures of Zak and Dakota" explaining cancer to children. Lead animator: Andrew Britton, Purdue University.*

High-fidelity mock-ups usually take a higher time investment, require less client imagination, communicate both form and function, and are more effective for persuading stakeholders. In the engineering professions, graphical 3D models are produced as virtual high-fidelity mock-ups on the computer using computer-aided design (CAD) software, based on the renderings provided. CAD designers usually create the 3D models. Engineering virtual high-fidelity mock-ups are produced using 3D computer programs to simulate how a product works. The process can be saved on a CD, as a video, or linked online and are produced by computer graphics technologists.

Here are some examples of high-fidelity mock-ups:

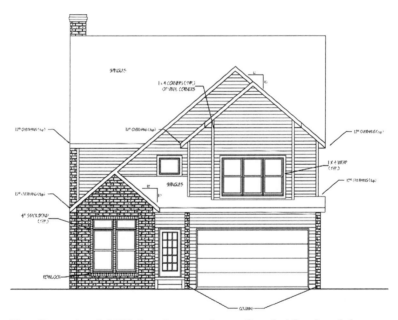

*Two-dimensional (2D) drawings can be produced either hand drawn or digitally drawn and used as mock-ups. Civil engineers use architects for the conceptual design. Drafters convert the designs of architects and engineers into technical drawings using software. This is a high-fidelity mock-up. House designed by Sharon L. Smith. Mentor: Professor Clark Cory, Purdue University.*

148  IDEA ENGINEERING

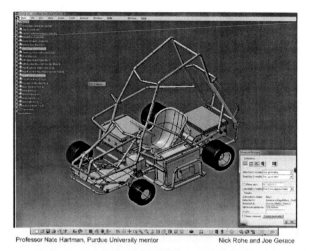

Professor Nate Hartman, Purdue University mentor — Nick Rohe and Joe Gerace

*Because of advances in technology, graphical 3D model high-fidelity mock-ups can be produced on computer-aided design (CAD) machines, based on the renderings provided. CAD designers usually create the 3D models using the software. The 3D computer applications have a testing component and allow mass properties analysis (i.e., mechanical properties, heat transfer, deformation, center of gravity, stress, mass, moment of inertia) to be tested on the screen. So, while the product is being designed, it can simultaneously be tested.*

*This is a high-fidelity dance studio mock-up designed by Chelsae Hooton. Mentor: Professor Clark Cory, Purdue University.*

## Prototypes

A real working model built for testing a new product, design, process, or concept is called a prototype. Prototypes are used in engineering design, electronics, software programming, interface design, semantics, and so on. A prototype gives the innovator an opportunity to evaluate whether the invention has any design flaws, and if it really works the way it was intended before the actual product, design, process, or concept is implemented commercially. The prototype is more detailed and polished than the mock-up. It involves a larger investment of time and money and is less disposable. Prototypes can be virtual models, physical to-scale models, or actual true-size models.

Before 1880, every innovator was required to present a prototype of the invention to the patent office when he applied for a patent application. When applying for a patent today, most are based on detailed drawings. The prototype can be used for demonstrations to license or sell an invention. In the United States, it used to be legal evidence that you were the first innovator with the "first to invent rule." A new law—the "first to file rule" took effect on March 18, 2013. This law is applied in every other country.[4]

### *Mechanical and Electronic Engineering Physical Prototype Models*

Engineering physical prototypes will be either scale models or actual true-size models. Models can be produced using wood, plastic, or casting metal by a professional prototyper. Because rapid prototype devices are now common, working parts can be made quickly without tooling. Mechanical and electronic engineers design devices that are mass-produced, and therefore producing a 3D solid model is one of the design cycle steps in prototyping.

*This is an actual true-size model of the Kart that is a working model. By Nick Rohe and Joe Gerace; Mentor: Dr. Nate Hartman, Purdue University.*

### 3D Printing Technology Prototypes

After two decades, we are now on the cusp of a 3D printing technology revolution. The additive manufacturing process in 3D printing technology was initially introduced to the prototyping world by producing scaled plastic models produced from 3D software for engineers. Many of the objects printed today are finished products, rather than only prototypes.

Artists are some of the first to experiment with the 3D printing technology. Industrial designers have fabricated toys, airplane parts, jewelry, and clothing (Cubify.com, 3DSystems.com). Designer Janne Kyttanen has a line of 3D-printed CubeX shoes, as well as iPhone and iPad cases and beautiful museum collection art.

As the 3D printing technology becomes more accessible, the opportunities for innovative ideas will increase. To illustrate how far the concept of 3D printing technology advanced, recently, it has been repurposed to print human cells and body parts.[5] Cutting-edge bio-engineering companies have fabricated human parts such as bone cartilage, human livers, and breast tissue from stem cells. The companies are seeking government funding and venture capitalists to realize their innovations.[6]

*3D-printed wrist splint prototype.*

### Chemical and Civil Engineers

As chemical engineers do not produce their products in large quantity, they have no need for designing a prototype. Chemical engineers build working small-scale functioning processes. One of the difficulties for chemical engineers is that the upscaling is not linear or simple. Much of this is trade secret, not patent.

Civil engineers also do not produce their products in large quantity; so they do not have a need for a prototype. The projects move directly to the construction stage. Because no prototype is needed for the civil engineering construction stage, engineers rely on design modifications. Modifications to design are found in all fields, and as-built versus as-designed are always issues in the field. In manufacturing, these are handled by change orders initiated by various stakeholders. They continue documentation called "as-built drawings," which document how the project was actually constructed and not how it was designed. In construction, the modifications are approved before they are made. The as-built drawings, which are prepared by the contractor, show the revisions to the original design using red ink and notes.[7]

### User Experience Design Virtual Models

In user experience (UX) design and animation, the virtual prototype—a high-fidelity screen rendering—is produced in the latter part of the design process. After the final revisions are made on the prototype, it becomes the actual product. The building of the prototype becomes the UX design or the animation.

*The National Pastime is an interactive game about the Japanese-Americans interned in camps during World War II. The graphics were done by Marcus Oania. The programming was done by Marin Petkov. Mentors: Dr. Bill Watson and Dr. La Verne Abe Harris, Purdue University.*

152　IDEA ENGINEERING

*These are prototype screen shots of an animated film "The Adventures of Zak and Dakota", explaining cancer to children. Lead animator: Andrew Britton, Purdue University.*

## Testing

Testing an innovative idea involves evaluating a specific product, process, or service in accordance with a set of standards. The **product**

**model** is tested to determine if what was delivered was what was promised, specifically: (1) customer service, (2) features and functions, and/or (3) interfaces. The traditional **business model** is tested to determine the validity of the economic model assumptions, specifically: (1) usage, (2) price, and (3) costs. The **marketing model** is tested to determine the effectiveness of the assumptions, specifically: (1) sales channels and (2) customer acquisition methods. The traditional business and marketing model testing is out of the realm of this book.

When an innovation is tested before being launched, it often comes in the form of a product model—alpha and beta testing. Alpha testing is an internal testing to validate the product, to confirm that the product works, and to identify any major flaws. When virtual engineering models are built on a computer, testing can be done interacting with the computer throughout the architectural process. The 3D computer applications have a testing component and allow mass properties analysis (i.e., mechanical properties, heat transfer, deformation, center of gravity, stress, mass, moment of inertia) to be tested on the screen for engineering products. The mock-ups and prototypes help gather functional requirements and features for the study.

Beta testing follows prototyping and is done close to the launch by external testers (customers and users). It takes place on the client side and involves testing the actual product in the user environment to determine whether the innovation delivers what was promised prior to its full deployment. This is done after the major flaws identified in the alpha testing are fixed. The purpose is to find final problems and defects and determine which defects will be addressed.[8]

The alpha–beta testing terminology originated from IBM for both software and hardware development; however, similar tests are used in manufacturing companies. In 2012, Toyota built 600 Prius plug-in demonstration automobiles for public beta testing. They sent them to large companies, the media, and fleets to test.[9]

| ALPHA TEST | BETA TEST |
|---|---|
| **Purpose of test** | |
| Ensures beta readiness by improving the quality of the product | Integrates user and customer feedback on the whole product, ensures release readiness, and improves the quality of the product in the field test |
| **Timing of test** | |
| When the product is in a near fully-usable state toward the end of the development | Right before launch |
| **Length of test** | |
| 3X to 5X the length of the beta test with many changes | A few weeks to a couple months with a few minor revisions |
| **Importance of test** | |
| Quality assurance and engineers looking for bugs, reviewing specs, etc. | Entire product team (quality assurance and engineering, marketing, support, etc.) looking for release of product |
| **Participants** | |
| Test engineers, product designers, employees, etc. | External real users and customers giving feedback |
| **Expectations of testers** | |
| Identifying lots of flaws, including major issues | Identifying fewer and minor problems |
| **Solving flaws** | |
| Critical issues fixed, some changes may occur resulting from early feedback | Critical changes are made; Feedback is collected for further versions of the product |
| **Achievements** | |
| Sets well-defined benchmarks and measures a product against them; methodology, efficiency and regiment | Explore the limits of a product in native environment; tests the reality of use |
| **Results** | |
| Beta ready? Does it meet design criteria? | An overall understanding of the customer experience |

Specific types of alpha–beta testing, such as in UX design, come in the form of Heuristic Evaluation,[10] Cognitive Walkthrough,[11] and usability testing (Exploratory Test,[12] Comparison Test,[13] Validation Test,[14] and Evaluation Test[15]). Usability evaluations for UX design are performed at different times in the design cycle, have different purposes, and have different approaches. These tests often measure ease of use, readability, navigation, visual appeal, and 508 compliance.[16]

## Food for Thought

Building the idea and testing it gives innovation validity. Now the innovation is ready to be introduced into the organization or society or both by the marketing team. From this point on, it is out of the hands of the engineering and technology innovator. This is one of the most exciting, yet stressful moments of the process.

# CHAPTER 10

# Worldviews

*A lot of people in our industry haven't had very diverse experiences. So they don't have enough dots to connect, and they end up with very linear solutions without a broad perspective on the problem. The broader one's understanding of the human experience, the better design we will have.*

—Steve Jobs

An idea may be a great one. The marketing department may have blasted the benefits of the product, process, or service throughout the company and around the world. But why are some great ideas adopted, whereas other great ideas go by the wayside? The answer is that it is not just about the idea. This chapter covers the social, economic, and political factors that may determine the fate of an innovation.

Gaining worldview knowledge helps engineers and technologists make responsible decisions that affect our society and the world in which we live. These values are intrinsically embedded in the decisions we make about whether or not to adopt innovation and adapt to change.

## Social Factors

Too many changes in an organization cause chaos, and not enough changes causes mediocrity. The organizational investment in producing an innovation or investing in the use of one must be balanced with the investment in people or the processes; otherwise, the innovation will not be successfully implemented. Often the driving factors of success are the human ones: dedication, commitment, wisdom to hire knowledgeable employees, creativity in finding resources, and the valuing of the innovation at all levels of stakeholders.

Internal conflicts of power set the stage for an innovation to fail and to be contrary to the strategic vision of the organizational leaders. An environment with task-oriented innovators, who are guided by champions who value them, sets the stage for innovation success.

Industry leaders should integrate those affected by the adoption of the innovation in the decision-making process by asking for input, valuing their knowledge and opinions, and implementing the best ideas. This empowers the workers. A healthy exchange of information and clarification of misinformation pull people together to form a cohesive unit. This in turn causes shifts in attitude and behavior. It aligns people and enables them to cope with change by working together toward a common vision. The level of participation of the stakeholders is an indication of the intensity of the buy-in for change and adoption of innovation. The measure of the stakeholder buy-in predicts the level of innovation adoption.

The social perceptions of workers who are adopting an innovation in their organization are of prime importance. When they perceive that the adoption of the innovation has more disadvantages than benefits for themselves personally and professionally as well as to the organization, or that the leaders do not have all the information needed to make a good decision, then change is resisted. The resistance to change is not necessarily a negative factor. It can act as a devil's advocate in determining the best decisions.

Industry managers should take some responsibility for the social perceptions of expectations, power, and conflict in their organization. American corporations today are overmanaged and underled. Successful leaders must be more than managers of organizational change. They must be champions of people.

Managers of industry are tasked with the responsibilities of planning and budgeting, hiring and firing of workers, and guaranteeing that the strategic plan will be implemented. They do this through the use of formal authority as well as through their ability to solve problems.

Unrealistic expectation or a complete lack of expectations confuses workers and causes excess stress and anxiety. Social perceptions are a real cause of organizational dysfunction and failed leadership, resulting in the creation of unproductive, frustrated, and apathetic workers. This in turn leads to an ineffective use of the innovation.

Leaders need to continuously monitor the social perceptions of the workers and keep them updated on the issues. Misunderstanding the purpose of the innovation adoption causes resistance, even when the intent is in the best interest of the individual and the organization. Often, leaders are not aware of the misunderstandings of the information. Many times the information is misconstrued, because it is never communicated in full to the employees. Rumors can run rampant when there is a lack of healthy dialogue between the management and the workers. Dispelling these misperceptions and educating the workers costs management time, money, and patience.

The most significant social factor of adopting innovation is perceived usability. Without the perceived value from humans, the innovation becomes useless. If people depend on the innovation to carry out work, if it maximizes efficiency, or if it increases the quality of the output, it has the potential of being adopted. In other words, it must be useful to organizations or individuals or both.

Workers who have a low tolerance for change often display fear and anxiety. This behavior wastes time and costs the company money in the long run. It takes the patience of a leader to dispel those fears and show support for the employee.

If workers perceive that the organizational change is too much too quickly, then it is up to the leaders to reassess the pace and merit of the adoption and prioritize the changes.

### *Social Example: Single-Serve Coffee Machines*

As the most significant social factor of adopting innovation is perceived usability, there are key lessons to be learned when introducing a new product into the marketplace, especially when the launch fails; take for example, the at-home brewing of coffee in a single-serve machine. Keurig is ranked the number one single-serve coffee machine in the United States. Consumers love the convenience.

When the patent for the coffee machine expired in September 2012, Green Mountain introduced the Vue V700 in February 2013. The consumer now had additional features, such as the ability to choose the coffee strength and the choice of temperature. The Vue V700 also allowed the option of making iced fruit-like drinks.

So why didn't consumers trade up? A selection of K-cup flavors for the Keurig machine was introduced frequently by Target and Peet's Coffee & Tea, whereas the selection of Vue pods were limited. In addition, the Vue V700 offered less product support to the consumers. It all came back to perceived usefulness.

### *Social Example: Videotex Technology*

Sometimes an innovation is not used to its potential. Some innovations have limited use, are only partly adopted, or are ahead of their time.

Videotex technology was an example of a technology that was only briefly adopted by society. It was ahead of its time socially and too ambitious for the existing technology available to support it. In the mid-1980s, I managed the computer graphic production of an international videotex venture called "Teleguide" in collaboration with McClatchy Newspapers of Toronto, Canada, and Phoenix Newspapers, Inc. The two-way interactive digital "information packages" were displayed on television monitors, which were placed in public kiosks. They provided free digital information for the general population, such as sports scores, weather, desert survival skills, higher education institution information, restaurant information, shopping mall product information, upcoming events in different cities, and so on. Some of the newsworthy information was written by the Teleguide staff. The other content was sponsored by "information providers"—the clients of the innovation. The kiosks were placed at the locations of the information providers as well as in other public places like shopping malls, universities, airports, the zoo, hotels, and so on.

The Phoenix Teleguide venture also consulted collaboratively with industry on several innovative, on-the-cutting-edge projects including the following: (1) a home banking project, (2) an in-hotel project, (3) and agricultural "e-mail" advice project for southern California farmers, who accessed information through their home computers. Remember, this was in the early to mid-1980s. The content, which was stored on mainframe computers, was similar in structure to what is known today as the World Wide Web. The graphics were originally created using vector graphics and AT&T Frame Creation Systems. Eventually, raster images were added. The end users made their selections from numeric pads on the kiosks. The digital information was shared internationally with end users in Canada, Japan, and several states in the United States, including Arizona, Florida, California, Nevada, and Hawaii. This graphic networking process predated the Internet's use of graphic websites.

Videotex was introduced into selected portions of society shortly before Apple Computer introduced the Macintosh in 1984. Computers were not commonly found in homes at that time, and people were not accustomed to retrieving information digitally. This innovation was introduced to society before people were ready to accept the change. One participant of the videotex venture said that the two-way interactive digital information would never be accepted, because he could not take it to the bathroom to read like his newspaper.

This technology was a precursor to web technologies. Many of us involved in two-way interactive ventures formed the groundwork of what was to become the World Wide Web. Nothing ventured, nothing gained.

## Economic Factors

The entrepreneur is not just interested in innovation for its own sake; innovation is important to the entrepreneur because it drives the economic engine of capitalism.

The new economy model suggests that it is cheaper to invest in innovation (i.e., technology) than in people. Many organizations make the mistake of investing more in the innovation than in training people in how the innovation should be used. Patterns of social behavior in an organization are changed because of the introduction of an innovation in the way work is done. Organizational change requires behavioral and skill modifications in the way that work is done, and this is frightening even for the workers who are used to technological change in their careers. When employees fear losing something of value, their morale is affected.

Adoption of innovation changes the economy. If the financial investment in purchasing or repairing the innovation is equal to or less than the revenue produced by using the invention, then the choice of adoption is good. Policy makers should not take sunk costs into consideration when deciding whether or not to move an innovation forward. Economists define sunk costs as those investments already incurred in the development of the innovation.

When the private sector is in an economically healthy situation, business investments in innovative products are not intertwined with the bottom line, especially if the product has a long technological life cycle. On a customer/user level, when the economy is healthy, people are more likely to personally invest in innovative products.

At the heart of the economic factors is the battle of engineering and technology industry standards. Perhaps the problem is not that there is no standard, but that often too many standards are introduced, and things get a little confusing.

### *Economic Example: The Drivers and the Copycats*

A small handful of companies are designated as consistent innovators; yet we hear very little about their competitors. We hear innovation success stories about Procter & Gamble, Target, and Apple, but we hear very little about the innovation success of their competitors: Unilever, Kmart, and Dell. Are the innovative companies particularly successful in their visionary executive leadership or in their R&D teams? Yes, to some degree, but these are not the key drivers. The key driver is sustained innovation over time through a culture of innovation that runs through every element of the organization.

The level of industry competition increases the opportunity for innovation, but it does not assure that the company will continue to focus on sustaining innovation. To remain competitive, companies cannot rely on playing the copycat game of follow-the-leader. This kills creative thinking and innovation skills. The only thing these copycats can do to keep their head above water is to get rid of costs and inefficiencies. Because copycats do not do their own research, they may end up duplicating a product with the same mistakes. They often do not understand what features and benefits the customers value.

### *Economic Example: VHS Versus Betamax*

Some great ideas are not adopted in lieu of not-nearly-as great ideas. The best choice of an invention is not necessarily the one that ends up being adopted. Often because of social, economic, and political factors, the lower level of innovation is chosen. It is often because more money is thrown in the direction of one of the products. In spite of the fact that one has to be of a certain age to have experienced the Video Home System (VHS) versus Beta tapes battle, I decided to include this example in my book. Some say it is an urban myth.

The family discussion began when my brother Glenn insisted that Sony's Betamax cassette was the superior technology versus JVC's VHS cassette. Glenn owned a Betamax unit and I owned a VHS unit. As time progressed, both technologies were purchased by consumers. Then a challenge was thrown into the competition. RCA Corporation wanted a device that could record three hours of the All-American game of football. The longest time recorded up to that point was one hour per Betamax cassette and two hours per VHS cassette.

JVC engineers took on the challenge. Betamax engineers were silent. Who knows if the JVC engineers really knew if they could pull this one off, but they spent six weeks developing a VHS tape with a four-hour capacity and were successful.

RCA ordered 55,000 VHS machines with up to a million more in the next three years. And to top it off, the VHS cassette was priced lower than the Betamax cassette.[1] This was an ugly battle worse than Blu-ray versus HD DVD.

I wish I had bet some money on this debate. When the purchase of the VHS technology surpassed the Betamax technology, I did the happy dance (it is a sibling thing). But Glenn did not concede this debate in my favor. He said that just because a technology is imbedded in a society, does not mean it is the superior technology. It means someone backed it up with money. Hmmmm. Rats! He was right.

### Academic Capitalism

The university is also affected by the economy. In the last quarter of the 1800s, universities shifted their focus from theology, education of men, and philosophy (all of which was overseen by clergy) to science-based disciplines such as engineering and technology. Dr. Sheila Slaughter, Professor at University of Georgia (formerly of the University of Arizona) and Dr. Gary Rhoades, my PhD Chair state that additionally there has been a "blurring of boundaries" between private and public sectors. "Academic capitalism" is discussed at length in the book *Academic Capitalism and the New Economy: Markets, State, and Higher Education* written by Dr. Slaughter and Dr. Rhoades.[2]

During the past 60 years, there have been many policy shifts in higher education. The most significant shift in the United States is the decrease of state allocations since the 1980s, which left universities with the responsibility of finding new ways of doing business, while facing dwindling resources. This national trend has resulted in higher education institutions, especially public universities, focusing less on knowledge for the public good, and more on being entrepreneurial to survive. In other words, universities have moved from an agency model to an enterprise model of investment, according to President Michael Crowe of Arizona State University.[3] The New American University should be a highly innovative venture that is open to opportunities for alternative means of funding through external agents.

In the new academic enterprise model, the university student is not only the consumer but the institution of higher learning also becomes the marketer. Once a student graduates, he or she is presented to industry as a product and potential contributor to the new economy. Higher education institutions have integrated into the new economy by networking and investing in the private sector to market products, processes, services, and institutions to students.

Arizona State University (ASU) College of Design's InnovationSpace is one such example.[4] When I met Professor Prasad Boradkar,[5] the director, I immediately connected with the way he ran his research and development laboratory based on integrated innovation[6] and solving problems in environmentally and socially responsible ways. In his laboratory, senior capstone teams composed of students from disciplines in engineering, industrial design, graphic design, and business collaborated on developing products that solve problems that are environmentally and socially responsible.

Each product is tied to an outside sponsor or partner. One sponsor that the InnovationSpace invested in is the Phoenix Fire Department. The team also worked in collaboration with the ASU Flexible Display Center to develop a prototype of a firefighters' helmet with a thermal-imaging camera installed to help identify victims during a fire. It was named Cyclops. This invention was streamlined for safety and eliminated carrying a bulky thermal-imaging camera into the fire.

Purdue University's Discovery Park is another example of an institution of higher learning connecting science, technology, and engineering with local, state, national, and international companies, corporations, and partners, such as Raytheon, Indiana Department of Transportation, U.S. Department of Agriculture, Nuclear Defense Group, and Sandia National Laboratories.[7] Discovery Park is a large-scale interdisciplinary research effort—a place for research and industry partnerships, entrepreneurial activities, and start-up offices to solve global problems and build a better world. Innovations are pushed to commercial development in an entrepreneurial environment. Since the 2001 inception, over 4,000 faculty members and students have participated, approximately 50 new start-up companies have been facilitated, and faculty and students have received entrepreneurial training. Discovery Park also links students working on projects to internships and job opportunities.

Research universities have transformed into the academic enterprise model to financially survive. On a national level, external resources must be increased to balance the decrease in state funding. Rhoades and Slaughter (1997) argue that one should follow the money to see where the priorities are placed in universities.

## Political Factors

Whether or not innovation is adopted in businesses or society depends on the support of the policy makers and their vision of the future and how the adoption of the invention can benefit the private sector or individuals or both. If it becomes too costly to repair a broken piece of the product, maintain it, or upgrade it, the chances of adoption are less likely. The most important factor, however, is whether all stakeholders—the policy makers, innovators, and customers/users—buy into the value of the invention.

The time spent on organizational change is usually longer than anticipated. Because of the emotional turmoil involved in implementing organizational change, individual employees often focus on their own agenda and self-interest as a survival mechanism, rather than that of the organization as a whole. This results in political behavior of manipulation, coercion, and concealed attempts to influence others, particularly by those who had formal or informal power in the organization.

### Political Example: Long Island Overpass Design

If this perspective is taken a step further, it is possible for engineers and technologists to design and construct something that gives authority, privilege, and power to a select group of people. A prime example is the Long Island overpass design by Robert Moses, a master builder from the 1920s to the 1970s. There is evidence that Moses had a predisposed political agenda when he designed the overpasses. Two hundred low-hanging overpasses were designed on the island with heights built to accommodate the middle and upper-middle class automobile-owning Caucasians on Long Island. Those who used the public transit—the African-Americans and the lower socioeconomic class—do not have access to the roads or Jones Beach, because the standard-sized 12-foot buses do not fit under the Long

Island overpasses.[8] This is how innovations can be manipulated to produce consequences directly affecting our lives.

*Political Example: The Atom Bomb*

Another example happened during World War II. It was the invention of the atomic bomb that caused social concern and led to the framing of ethical choices about science and technology coming to the forefront. Before that time, engineers, scientists, and technologists traditionally had the charge of solving problems, not questioning the ethics of the problems. The social, economic, and political responsibilities resulting from the atomic bomb reinvented how engineers, scientists, and technologists thought about innovation. They needed to ask themselves, "Just because it could be invented, should it be?"

It has taken over half a century to see that one of the casualties of war is unveiling the truth. The declassification of U.S. government secret files revealed that the bombings of Hiroshima and Nagasaki were political decisions, not military ones. President Truman needed to destroy a dense urban population in Japan as a power play with the Soviets, and in particular with Joseph Stalin. By politically framing the atomic bomb, Truman's goal was to keep the Soviets from entering the war, and accepting conditional surrender of the Japanese.[9]

# High- and Low-Impact Innovation

When an innovation changes our behavior, our social habits, and our priorities, it is a high-impact innovation and has economic and behavioral ramifications on society. It is useful to our personal and business lives. High-impact innovation may provide business opportunities and a faster return on investment. It may impact how we communicate with others.

Because consumers/users expect things to work right away and we have high expectations, usually the result is that the demand and the customer morale drops shortly after the release, until either user expectations are adjusted or additional improvements are made to build customer satisfaction. When I worked in industry, the first innovation to market (bugs and all) typically won the brass ring. Any minor imperfections were worked out and delivered after the market release.

It takes time for a radical and novel invention to be assimilated into society. People are slow to accept a new innovation. Adoption means change and substituting the old way.

The same is true for higher education. Acquiring knowledge and basic research at universities requires sponsorships and is slow in achieving results; however, these innovations can change the world. That is why collaboration between the private sector and academia is necessary. The innovation must be affordable, have clear benefits in the lives of the users, and have effective marketing to increase awareness.

When an existing product, process, or service is improved, it is called a low-impact innovation. As we have discussed, the majority of innovation falls into this category. Low-impact or incremental innovation moves slower because its purpose is to fine-tune an existing product, process, or service. This may have a high impact on customer satisfaction and a faster return on investment at the release.

## Food for Thought

Decisions that are made about innovation can have underlying social, economic, and political effects on our workplace and world; so engineers and technologists need to be responsible for these ramifications. When engineering and technology innovators are designing the newest product, process, or service, they should take a moment to think about how these decisions will affect different genders, ages, nationalities, or those with a different socioeconomic status.

# CHAPTER 11

# Final Thoughts

*I've gained lots of knowledge. I now know a thousand things that won't work.*
—Thomas Edison, after 1,000 attempts to find the perfect filament for the incandescent lamp, went on to try 800 more ideas, until he found the best solution.

Conversations I have had about innovation revolve around collaboration, ideation, implementation, and value creation. During one of those discussions, my colleague and friend, Dr. Jon M. Duff, said something I wish I had thought of:

Philosophers explain WHY the universe exists (and the output is religion or philosophy).

Scientists explain HOW the universe behaves (and the output is scientific theory).

Engineers CHANGE how the universe behaves (and the output is technology).

Technologists INTEGRATE technology into society (and the output is training, implementation, evaluation, etc.).

Innovation is different with each group.
—Dr. Jon M. Duff

In the journey you have traveled with me, we have reviewed three sections of creative thinking and innovation: (1) the foundation, (2) thinking and doing, and (3) insights. Our conversations on ways of getting your mind right to prepare for the adventure, the lives and traits of visionaries and innovators, an overview of creative thinking and innovation, roadblocks, and models of innovation and idea teams have given you a foundation on the topic of idea engineering.

Creative thinking and innovation is about learning different ways of thinking and doing things. It begins by identifying the problem, which then opens up the art of thinking like a designer, inventor, and entrepreneur. With the plethora of ideas, one idea is selected to move forward to be built and tested.

Insights on our discussion are touched on in Chapter 10 "Worldviews." It takes a review of worldviews to gain an understanding of why certain ideas are adopted in our society. In Chapter 11, I end our conversation for now with things to think about as you go about your work and play, and I share with you the place I am coming from at this point and time in my life.

If you are going to be inspired to be a creative thinker and innovator, you need more than just academic knowledge. You need chutzpa (fearlessness) and persistence, respect for the failures and successes of past innovations, and the ability to apply what you have learned. This involves change. There are three basic responses to the fear of change. You either fight, flee, or freeze. Freezing—doing nothing—is a choice. Usually people freeze, because they wait until the problem goes away, or they wait until everything is perfect.

> Too many people spend too much time trying to perfect something before they actually do it. Instead of waiting for perfection, run with what you've got, and fix it along the way.
> —Paul Arden

Innovation in industry requires trained professionals such as engineers, technologists, scientists, and business people. This book has been about applying your newly gained creative thinking and innovation knowledge for the future.

## The Future of Idea Engineering

The question that keeps playing in my head is "Have we as a country lost the excitement of discovery?" I long for the day that innovators and their breakthrough creations are celebrated in our society, and creative thinking and innovation are universal educational topics that are valued.

An invention that comes to mind is the lithium-ion battery, in which the prototype was invented 30 years ago. It is used to power electric cars, GPS navigational systems, tablet computers, and military aircraft drones. But who can tell me the name of the innovator?[1]

### *Improving Creative Thinking and Innovation in Industry*

One of the purposes of this book is to improve the approach of idea engineering as it exists in the private sector today. Years ago, when I was at Arizona State University, I interviewed senior engineers and managers as a part of my research. When I asked what characteristics they were looking for when hiring engineers and technologists, they said they wanted creative thinkers with inquisitive minds who were not afraid to break boundaries. They wished that creative thinking and innovation was taught as a part of the engineering and technology curriculum at research universities. The senior engineers and managers also said that they wished they had a resource on creative thinking and innovation to put some life into the employees who were static and stuck in status quo. I am hoping the contents of this book can help.

### *Improving Education*

The American culture that values education needs the partnership of the private sector. U.S. companies need to invest in STEM (Science, Technology, Engineering, and Math) beginning in the kindergarten through 12th grade educational system. That means that companies should follow the examples of Google, who inspire K-12 innovators by launching a global science fair each year, and NASA, who continue partnering with the National Science Teachers Association.

Creative thinking and innovation needs to become a required part of the engineering and technology curriculum in higher education to produce a new breed of exemplar engineering and technology innovators who are global leaders. This will entail offering a course on idea engineering or incorporating the concepts into other courses. It will also involve a shift to more multidisciplinary curriculum including business, psychology, design, and so on. Engineers and technologists of the future must not just invent technological devices. They must be in touch with the social, economic, cultural, and political consequences of their choices.[2]

The alliance formed by industry and academia will help budding innovators understand the desirability, feasibility, and viability of innovation. Many engineering and technology education disciplines include a senior project for undergraduate students. Many private sector organizations partner with universities to collaborate on product development by submitting market challenges for the students. Some are even funded. This collaboration can lead to future employment as well as possible patents. At project completion, the students present their project to the corporation as well as to university personnel and students. These seeds of innovation have all the potential of growing into a marketable product, process, or service.

## Serendipity and Defining Moments

So much can be learned by studying case studies of innovation, whether it be in engineering, technology, or the sciences. Serendipity plays a role in the innovation process—finding worth in the unexpected. Innovation does not always follow a clean linear path. Sometimes an idea is developed to solve one problem but actually ends up solving another. Remember the 3M story of Post-it notes? The role of luck and timing has much to do with the success of innovation.

I want to leave you with a final story about innovation. Our story begins in the year 2000. President Bill Clinton announced on June 26 that J. Craig Venter, a visionary biologist and founder of a company called Celera, and Dr. Francis Collins, leader of a well-funded project called "Human Genome Project," had both succeeded in mapping out the entire human genome. This genetic map was to be the key to unlocking causes of diseases such as cancers. This was a defining moment.

The door for medical innovation was wide open. The unlocking of the human genome enabled a small team of Celera researchers to discover a way to inhibit a pathway they believed would help autoimmune diseases such as rheumatoid arthritis five years later (2005). In a small pharmaceutical laboratory, they experimented with designing and synthesizing molecules, which resulted in the discovery of an experimental drug I will call Drug A.[3] This was another defining moment. Drug A was a drug waiting for a purpose. Experimentation requires risk with no guarantees that a great idea will turn into revenue. Money was tight for the company. A few months after the drug discovery, the company decided to change direction and shut down its pharmaceutical program. The medical research team members were laid off, and were spread across the world (i.e., China, Australia, and the National Cancer Institute). The experimental Drug A was archived, just like many other great ideas. Fortunately, Celera did not throw away their idea; otherwise, our story would have a different ending.[4]

As we have discussed, the collaboration between a small group of people from research universities and business organizations brings together great minds from different places. The setting can be the most innovative, as opposed to working alone.

About the same time that the Celera researchers burned the midnight oil in their laboratory, researchers at Stanford University in Dr. Ronald Levy's blood cancer lab were collaborating with a San Francisco Bay area biotechnology company[5] using another experimental drug I will call Drug B.[6] Some results were promising, but most were not good. Two entities (Celera and Stanford University) existed in a parallel universe trying to solve a problem.

Dr. Richard Miller, an oncologist, was a frequent visitor of the Stanford University laboratory. Dr. Miller was a Bay area entrepreneur and CEO of Pharmacyclics, a struggling drug development company. Dr. Miller began to visualize the possibilities of experimenting with the archived Drug A and made an offer to Celera. After several failures and FDA disputes, Dr. Miller finally purchased the drug in 2006 from Celera for almost nothing.

Innovation comes with taking risks. Dr. Miller's research team substituted Drug B in the blood cancer experiment and replaced it with the recently purchased Drug A, experimenting on dogs with lymphomas with more promising results.

Often it takes courage to value clinical need over conventional wisdom, and it takes a great deal of luck. Dr. Miller wanted to pursue a less traditional approach to speed up the path to a human clinical trial for blood cancer; so he looked to the money. The question was whether it was worth a $1 million investment to move forward on testing Drug A on humans with blood cancer. Dr. Miller's answer came when he told his colleagues, "I have patients who are dying," who are out of options. He believed in Drug A and risked moving forward on the human clinical trial. As of the writing of this book (2014), Drug A is in phase 3 human clinical trial and will be FDA-approved within the year. It ended up helping patients with leukemia. It is not a "magic bullet," but it keeps leukemia at bay.

Sadly, Dr. Miller was not able to triumph in his decision. In 2008, Chairman of the Board Robert Duggan (Scientology's biggest donor) fired Dr. Miller for differences of opinion.

And our story continues with another serendipity moment. The ex-Celera research team members found out by accident (by a colleague's online search ending up at Dr. Brian Koffman's blog) that their Drug A innovation was successfully being used in clinical trials throughout the United States to help leukemia patients. They were overjoyed at the news.

## Food for Thought

Today, I am one of the white lab mice in the clinical trial with Drug A. I am filled with gratitude for all the inventive stakeholders, who are a part of the journey to bring this cancer drug to market. I had to leave Purdue University at the height of my academic career, because I was diagnosed with blood cancer and was symptomatic. When I was labeled a "high-risk cancer patient" and given a poor prognosis with no avenue for successful treatment, I put on my creative thinking hat and began looking for the top cancer experts and researchers in the United States. At the time when I did my homework, I was discouraged with the small amount of research that had been done on the blood cancer I had. I formed my personal medical team made up of experts from M.D. Anderson in Houston (Dr. Michael Keating), the National Institutes of Health near Washington, D.C. (Dr. Mohammed Farooqui, Dr. Adrian Wiestner, Dr. Georg Aue), and a wonderful local hematologist/oncologist (Dr. Robin Obenchain). Writing this book has offered me an amazing diversion and an opportunity to give what hopefully is not my last lecture.

With gratitude & love,<br>
Dr. La Verne

# About the Author

*Remembering that I'll be dead soon is the most important tool I've ever encountered to help me make the big choices in life. Because almost everything—all external expectations, all pride, all fear of embarrassment or failure—these things just fall away in the face of death, leaving only what is truly important.*

—Steve Jobs

## Industry Biography

Before my venture as a university professor, I was an industry manager and director in the roles of art director, creative director and production manager. Even today, you cannot be shy, if you are a minority female entrusted to perform successfully in these leadership positions. Contrary to common belief, I really was a quiet child growing up, until I came upon my real self at about age sixteen. I had to learn to be assertive. I always joked that I was Japanese until the German side of me came out. LOL.

Taking on a leadership role in industry was a significant achievement at the time for me—being a minority female in a male-dominated field.

One of my highlights was when I had the opportunity to manage a team of artists, writers, and technologists in a cutting-edge experimental digital technology in a joint venture with McClatchy Newspapers of Toronto, in collaboration with Canada, Japan, and states such as Arizona, California, Nevada, Florida, and Hawaii. I led my department through the transition from analog to digital processes.

I also had the privilege of being a beta tester for Adobe software products, and working in advertising and the newsroom. The common thread in my industrial leadership experience was my involvement in the research and development of innovative and emerging technologies to solve information problems.

## Academic Bio

My love for art, science, writing, and emerging technologies, motivated me to pursue higher education as a career after I left the private sector, and focus my academic research and activities in teaching people how to think creatively. I am grateful to have had the opportunities to participate in scholarly activities and teach at Purdue University as a tenured Associate Professor of Computer Graphics Technology. I am also grateful for my experience at Arizona State University (A.S.U.) as an Assistant Professor of Graphic Information Technology.

One of my fondest accomplishments is being the founder of the IDEA Laboratory, which is a creative thinking, interactive media, and animation research and development laboratory. This laboratory began at A.S.U. and continued successfully at Purdue University in West Lafayette, Indiana with my colleague Professor Nicoletta Adamo-Villani. The mission of the laboratory is to apply new models of creative and innovative thinking that seek to produce interactive media and animation products and processes to improve the lives of people in our society.

When I first began putting my lab together, I was told by an academic person to never hire undergraduates to do research. I did not listen, because when I see talent, I do not care whether a person is a rookie or a highly-experienced researcher. As long as you deliver, I will support you with funding. The same goes for the business world. My incredible students that I named "IDEA Gurus" not only delivered for me, but made my experience in academia purely delightful. That is why I dedicated this book to them.

When I announced that I was interviewing students for the positions in the lab at Purdue University, many were excited that some of the projects involved developing video games. Boy, were they surprised when they found out that what we were doing was developing "serious games" that were educational and helped make society better – and they did not include decapitating the enemy! Video games are a powerful and captivating form of media and have great appeal to younger generations. Computer games have such a potential to educate because they can motivate players through entertainment. However, unlike traditional video games, serious games focus on the ability of the user to retain information through engagement and fun.

An example of one of the projects we storyboarded, produced, and delivered was an online video game about El Dios Verde (the Green God) and the Virus Warriors.[1] It was produced in both English and Spanish. The IDEA Gurus and I worked with philanthropists like my friend Sally Russ, physicians in Houston and the Yucatan peninsula, along with the TEAMM of Brazos Abiertos,[2] a non-profit group from Houston and the Yucatan Peninsula, and a group of high school students in the Yucatan, who wanted to learn about human immunodeficiency virus (H.I.V.), since it was spreading at a rampant rate in the peninsula. The video game was installed in a local high school computer laboratory, and in the cyber cafés that a local priest had constructed in four third-world villages in the Yucatan. I am really happy that this online video game will help in a small way to make the world and the Yucatan society just a little better.

Several other serious games that were designed and developed in the IDEA Laboratory included a game that helped children deal with the effects of living with spina bifida,[3] one that taught high school students about citizenship through learning about Japanese-Americans in internment camps during World War II,[4] and an interactive game to teach deaf children about mathematics.[5]

We designed information for interactive science, technology, engineering, and mathematics (STEM)-related research. Funded grants from the government and industry enabled the research teams to use problem-based learning and interactive media to remove barriers to success for under-represented minorities and women in engineering and technology,[6] and enhance visualization skills for engineers and technologists.[7]

Some of our IDEA Laboratory e-projects involved developing lectures for online education, as well as outside workshops on the fundamentals of business.[8] Another project was about haptic learning[9] and three dimensional environments.[10] We even did a rap video to promote one of the summer camp events in the Veterinary Medicine Department,[11] and we produced a 3D animation about how children deal with cancer through identifying with turtle and bear characters.[12] There were a number of them I had to pass on to other colleagues. What an incredible ride!

And since I am an academic, and publish or perish is my motto, some of my presentation and published research avenues are: Delmar Cengage Learning (*Visualization, Modeling, and Graphics for Engineering Design*), the National Collegiate Inventors & Innovators Alliance (NCIIA), ACM SIGGRAPH, American Society for Engineering Education, Communications Journal, National Association of Industrial Technology, Engineering Design Graphics Journal, and Institute of Behavioral and Applied Management (IBAM).

Along with my published refereed journal articles, conference proceeding publications, white papers, book chapters, illustrated books, international and national presentations, I have had the wonderful opportunity to act as the Director of Publishing and Editor of *The Engineering Design Graphics Journal*, an internationally-respected professional journal. I have had the opportunity to be a reviewer for a number of academic journals, such as *The Journal of Technology*, *The International Journal of Geometry and Graphics*, and *The International Journal of Knowledge, Culture, and Change Management*. I have also been a reviewer for conference papers for the American Society for Engineering Education (ASEE), and Frontiers in Education (FIE).

# Notes

## Love & Gratitude (Acknowledgments)

1. Ten Golden Rules by Rocky Harris:
   1. Come up with solutions to problems before you ask for help.
   2. Take your work very seriously, but never ever take yourself seriously.
   3. Don't sweat the small stuff. If you aren't making mistakes, you probably don't have enough to do. Learn from mistakes and move on.
   4. Reinforce the importance of planning. A lack of planning by others should never result in an emergency for you.
   5. Reciprocity: Treat everyone, including interns and janitorial staff, the way you expect to be treated.
   6. Smile often. You didn't pursue a career in sports (technology, engineering, marketing, etc.) to be miserable.
   7. Care for others, and worry about yourself: If you are caught up with what everyone around you IS or ISN'T doing, you've already failed.
   8. Believe in your leadership team: Don't second-guess everything they do and say.
   9. Live every moment as if it is your last. Wake up everyone morning determined to make a difference, to make yourself and everyone around you better.
   10. Everyone is replaceable. As a mentor of mine, Bill Walsh said, "The sign of a good leader is someone who moves on and their absence goes unnoticed, because they set the organization up for success long after their departure."

## Chapter 1

1. Pausch and Zaslow (2008).
2. MacKay (2012).

3. Hayes and Comer (2010).
4. Butler (2002).
5. Hindle (2008).
6. Black and white thinkers think in clear and concrete terms. They like rules that do not change. They do not like abstract and ambiguous thinking.
7. Allen (2003).
8. Showkeir and Showkeir (2013).
9. Vishwanath (2013).
10. Byrne (2007).

# Chapter 2

1. http://www.rubegoldberg.com/
2. http://www.youtube.com/watch?v=8X7f2zdQ3hc
3. http://www.youtube.com/watch?v=xdPDn1KUz_A
4. http://www.purdue.edu/newsroom/rubegoldberg/index.html
5. Pasachoff (1996).
6. Bellis (n.d.).
7. http://www.examiner.com/article/thomas-edison-s-recipe-of-adhd-a-tremendous-ability-an-un-tremendous-world
8. Edison (2013).
9. Einstein (2013).
10. Piccioni (2011).
11. Lampton (n.d.).
12. http://www.da-vinci-inventions.com/robotic-knight.aspx
13. http://science1.nasa.gov/science-news/science-at-nasa/2005/04oct_leonardo/
14. Bath (2013).
15. Zuckerberg (2013).
16. Bezos (2013).
17. The cloud is a data storage service that requires special software (Amazon's MP3 Uploader and Downloader) and is available on the web and on Android but not on iOS. As of the writing of this book, Amazon has blocked streaming through Safari, and the Amazon MP3 app doesn't exist in the App Store. Oh, here we go again... the political framing of technology.

18. http://www.businessinsider.com/everything-you-need-to-know-about-jeff-bezos-amazing-10000-year-clock-2013-8
19. Gates (2013).
20. Isaacson (2011).

# Chapter 3

1. http://www.searchquotes.com/search/art+begins+in+imitation/
2. Divergent thinking is a spontaneous, free-flowing, randomized method of generating ideas by exploring many possible solutions.
3. Subjective associative thinking is a process of relating or connecting ideas existing in the mind.
4. Lateral thinking is a term coined by Edward de Bono in 1967. It is a method of problem solving using a creative and indirect approach that is not often obtained by a traditional sequential and logical method.
5. Sloane (2012).
6. Hernandez and Varkey (2008).
7. Report to the President (2012).
8. Lohr (2009).
9. Donofrio (2004).
   *Note:* Australia, the Europian Union, and Canada had begun the implementation of innovation skills policy.
10. Kim and Mauborgne (2005).

# Chapter 4

1. Hare (2013).
2. Appleseed (2013).
3. Levy (2000).
4. Poeter (2011).
5. 3M (2002).
6. Kelly and Littman (2001).
7. Isaacson (2011).
8. HP (n.d.).
9. Cerf (2011).
10. Harris (2004).

11. de Bono (1999).
12. Walters (2008).
13. Collins and Porras (1997).
14. The Incentive Federation Inc. (n.d.).
15. SHRM (2012).
16. Hymowitz and Murray (1999).
17. Anders (2002).
18. http://www.1000ventures.com/business_guide/cs_efficiency_canon_ps.html
19. Edmunds (2013).
20. Deschamps (2008).
21. Harris (2004).
22. Harris (2006a) and Harris (2006b).
23. The professor's name has been changed to protect the guilty colleagues.
24. Harris (2004).
25. A "doubting Thomas" is a term for a skeptic who refuses to believe unless he has a personal experience. It is a person who is lacking in faith. This is a reference from the Apostle Thomas.
26. de Bono (1992).

# Chapter 5

1. Roberts (2011).
2. Roberts (2011).
3. The Bridge to the Future website (2012).
4. disneyinstitute.com (n.d.).
5. Isaacson (2011).
6. Roberts (1998).
7. Roberts (1998).
8. Albrecht (2013).
9. Rother (2010).
10. Rother (2010).
11. Iverson (1998).
12. Bloomberg Businessweek Magazine website (2006).
13. Carr (2013).
14. Power (2013).

15. Alcoa (n.d.).
16. Pitney Bowes (n.d.-c).
17. Li and Solis (2013).
18. Roberts (1998).
19. House and Price (2009).
20. Bloomberg (2010).
21. IDEAlliance conference (2006).
22. Chesbrough (2003).
23. Alcoa (n.d.).
24. McKinney (2012).
25. Gratitude to Magdalena Soto and Divya Marwaha, two of my ASU IDEA Gurus.
26. Pitney Bowes (n.d.-a).
27. Pitney Bowes (n.d.-b).
28. Pitney Bowes (n.d.-c).
29. Collective intelligence is a sociological phenomenon where a shared or group intelligence emerges from the competition and collaboration of many people.
30. King (2010).
31. All expenses in a zero-based model budget must be justified for each new fiscal period. It starts from a "zero base," and every project within a company is analyzed for its costs and needs. A new budget is then built around the upcoming period, regardless of whether the budget is higher or lower than the previous one.

## Chapter 6

1. Quote by Einstein: "If I had an hour to solve a problem I'd spend 55 minutes thinking about the problem and 5 minutes thinking about solutions." (http://www.goodreads.com/quotes/60780-if-i-had-an-hour-to-solve-a-problem-i-d)
2. Isaacson (2011).
3. Ishikawa (1991).
4. Novack and Cañas (2006).
5. Buzan (1983).
6. Gore (2013).

7. Gates (2005).
8. Mariaraj (2005).
9. This problem statement is from Dr. Nathan E. Bench's Purdue University dissertation: "Spatial-Haptic Perception of Virtual Shape Environments."
10. Rother (2010).
11. Thomas (2013).
12. Rother (2010).
13. Thomas (2013).
14. Thomas (2013).

# Chapter 7

1. The dobro is a resonator slide guitar that is usually played on one's lap.
2. A capo is a device used on the neck of a fretted stringed instrument to raise the pitch.
3. Charyton and Snelbecker (2007).
4. Boches (2012).
5. Michalko (2011); http://creativethinking.net/DT13_TrueAndFalse.htm?Entry=Good
6. Cherry (n.d.).
7. YouTube.com TED talks.
8. Osborn (1979).
9. Leifer (1998).
10. Kelly and Littman (2001).
11. Harris (2005a) and Harris (2005b).
12. Walters (2008).
13. de Bono (1990a) and de Bono (1990b).
14. de Bono (2005).
15. Brainwriting.
16. Berman (2006).
17. Schwartz (2009).
18. Brainstorming.co.uk (2006).
19. Nolan and Williams (2010).
20. Terniko, Zusman, and Zlotin (1997).

21. Kaplan (1997).
22. http://www.ers.usda.gov/publications/ap-administrative-publication/ap-037.aspx
23. http://www.innovationinpractice.com/innovation_in_practice/closed-world-principle/

## Chapter 8

1. Section 508 of the June 2001 Rehabilitation Act, often referred to as "508 compliance," requires that all electronic information used by the federal government be accessible to employees and the public with disabilities. What this means, for example, is that if a person is blind, the website must have an audio feature so that the content can be read out loud.
2. Sadowski and Sorby (2012).
3. Linstone and Turoff (2002).

## Chapter 9

1. Buxton (2007).
2. User experience (UX) design. User experience design is an emerging discipline and a profession encompassing information design, interaction design, websites, or software applications that involve the human element (the user) in order to work. UX design draws from cognitive psychology, information architecture, storytelling, haptics, environmental design, and so on.
3. animated.gif
4. http://gigaom.com/2013/03/18/first-to-file-patent-law-starts-today-what-it-means-in-plain-english/
5. Smith (2013); Leckart (2013).
6. Jeffery (2013).
7. AIA Knowledge Resources Team (2007).
8. Some software products seem to be in perpetual beta mode (i.e., Gmail). My take is that software development is never quite done.

9. http://www.thetruthaboutcars.com/2012/09/review-2012-toyota-prius-plug-in-hybrid/
10. A Heuristic Evaluation is a holistic usability review with usually two to three experts. This is often done before any of the other usability tests to work out any major bugs. But because the most important reviewer is the user of the design, the Heuristic Test is never used as the sole usability test.
11. A Cognitive Walkthrough is done to prepare for the prototype test during the early design phases before coding has begun. Before the product, process, or service is finalized, expert users evaluate it going through the latest version with the mindset of the actual users to identify any missed problems or difficulties in interactivity. This stage is a way to evaluate the prototype to observe if the interface actually does what is expected. Does it meet the user's/customer's needs? Any remaining problems are then identified and corrected.
12. An Exploratory Test is done in the early part of the design cycle, and it does two things: (1) It determines if the interactive product, website, or software features are useful and appropriate and (2) it allows the user to explore the functions to determine if the interactive product, website, or software does what he or she expected as well as point out what part of the design the participant found to be confusing.
13. The Comparison Test can be performed at any time, but is usually most beneficial to the designer when it is done early in the design cycle. The test participant (user) is given a list of tasks to perform using two or more very different designs. The end result is that the best design is selected, or the strongest features of each product are combined to create a new design.
14. The Assessment or Validation Test, which is the most common type of usability test, is usually given during the mid-level part of the design cycle, when the navigation has been decided upon. The test monitor determines how well the features have been implemented and understood by the participant through measuring the completion of a list of tasks. Sometimes a video is recorded to review and measure the responses of the user, instead of using a human monitor. Time limitations for each task are often noted as a benchmark of completion without the knowledge of the test participant.

15. The Evaluation Test is given after the UX design product is live and in use. It measures the completion of tasks by the participant to determine if the product is meeting the goals, and if the revisions have increased the usability.
16. Section 508 (June 2001 Rehabilitation Act) is a piece of legislature that requires all digital information (including websites) used by the federal government to be accessible to disabled individuals.

## Chapter 10

1. Rothman (2009).
2. Slaughter and Rhoades (2004).
3. Crow (2011).
4. http://innovationspace.asu.edu/
5. I am so very grateful for Professor Prasad Boradkar. He was my inspiration in creating my IDEA Laboratory, both at Arizona State University and Purdue. And gratitude to Professor Jacques Giard from Arizona State University, who spent many hours mentoring me.
6. Integrated innovation asks four key questions: What is desirable? What is good? What is valuable? What is possible?
7. http://www.purdue.edu/discoverypark/
8. Robert Moses.
9. Black (2002).

## Chapter 11

1. Professor John Goodenough (born of American parents) and his research team were at Oxford University when they developed the first experimental lithium-ion battery prototype. Professor Goodenough is now teaching mechanical engineering and materials science at the University of Texas in Austin. In June 2012, John Goodenough, Rachid Yazami, and Akira Yoshino received the 2012 IEEE Medal for Environmental and Safety Technologies for developing the lithium-ion battery.

   The plastic, rechargeable lithium-ion battery was invented in 1995 by a team of Bellcore researchers led by Jean-Marie Tarascon, who spent most of his career in the United States.

2. National Academy of Engineering (2005).
3. Ibrutinib, formally known as PCI-32765.
4. 3M (2002) archives all their ideas for future possibilities.
5. Rigel, a Bay area biotechnology company.
6. Fostamitinib.

## About the Author

1. IDEA Gurus: Brian McCreight, Marcus Oania, Jacob Brown, Michael Hoerter, Marin Petkov, John Williford; Faculty director: Dr. La Verne Abe Harris.
2. TEAMM stands for "Teenage Education on AIDS in Mérida, México" http://www.hivyucatan.org/en/
3. Samantha's Song Game. IDEA Guru: Hana AlJaberi; Faculty director: Dr. La Verne Abe Harris.
4. National Pastime Game. IDEA Gurus: Marin Petkov, Marcus Oania; Faculty director: Dr. Bill Watson and Dr. La Verne Abe Harris.
5. Mathsigner™. IDEA Guru: Jason Lestina; Faculty directors: Dr. Ronnie B. Wilber and Professor Nicoletta Adamo-Villani.
6. The HerStories video vignettes, which were produced by the IDEA Laboratory, contributed to the NSF-funded CareerWise web tool to help women persist in STEM doctoral programs. IDEA Guru videographer: Kenny Yanga; Video director: Dr. La Verne Abe Harris; Project directors: Dr. Mary Anderson-Rowland and Dr. Bianca Berstein, Arizona State University.
7. EnVISIONS: Enhancing Visualization Skills: Improving Options and Success was an NSF-funded project representing the partnership of seven universities. Faculty team at Purdue: Dr. La Verne Abe Harris, Dr. Kara Harris, Dr. Mary Sadowski and Dr. Patrick Connolly.
8. IDEA Gurus: Kenny Yanga, Marcos Lujan, Jon Garza, DeAnna Lara, Magdalena Soto, Divya Marwaha, Dustin Watson, Jaret Lynch and Adam Lessell.
9. IDEA Guru: Estaban Garcia
10. 21st Century World is an online 3D interactive environment that can be experienced as a portal for student and industry professionals

to explore and learn about what types of enhancements are possible with nanotechnology. IDEA Gurus: Eric Johnson, Tyler Penrod, Marin Petkov and Matt Barton, Faculty director: Professor Nicoletta Adamo-Villani.
11. Boiler Vet Camp music video. IDEA Gurus: Project manager: John Kessler; Music composer: Jacob S. Brown; Storyboards: Marcus Oania; Lyrics and vocals: Alesha Peterson, Jimmy O'Connor and John Lindstrom; Faculty director: Dr. La Verne Abe Harris.
12. The Adventures of Zak and Dakota is a cancer educational animation for young children. IDEA Guru animator lead: Andrew Britton; Collaborators: Purdue University IDEA Laboratory, Boston University and Children's Hospital of Boston; Faculty director: Dr. La Verne Abe Harris.

# Bibliography

3M. (2002). *A century of innovation: The 3M story.* Saint Paul, MN: 3M Company.

AIA Knowledge Resources Team. (2007). *Terminology: As-built drawings, record drawings, measured drawings.* Retrieved June 1, 2013, from http://www.aia.org/aiaucmp/groups/ek_members/documents/document/aiap026835.pdf

Albrecht, K. (2013). *The world's billionaires: Jeff Bezos.* Retrieved April 15, 2013, from Forbes: http://www.forbes.com/profile/jeff-bezos/

Alcoa. (n.d.). *Alcoa and DuPont collaborate on new hurricane-resistant architectural panel system.* Retrieved March 1, 2013, from Alcoa website: http://www.alcoa.com/building/en/news/releases/alcoa_dupoint.asp

Allen, C. (2003). *The benefits of meditation.* Retrieved February 1, 2013, from Psychology Today: http://www.psychologytoday.com/articles/200304/the-benefits-meditation

Anders, G. (2002, February). How Intel puts innovation inside. *Fast Company* (56).

Appleseed, B. (2003). *Steve Jobs' interview at 60 minutes (2003).* Retrieved February 15, 2013, from http://everystevejobsvideo.com/steve-jobs-interview-at-60-minutes-2003/

Bath, P. (2013). *Biography.* Retrieved January 27, 2013, from http://www.biography.com/people/patricia-bath-21038525

Bellis, M. (n.d.). *The life of Thomas Edison: Family background, early years, first jobs.* Retrieved June 15, 2013, from: About.com: http://inventors.about.com/od/estartinventors/a/Thomas_Edison.htm

Berman, T. (2006). *Six degrees of separation: Fact or fiction?* Retrieved March 15, 2013, from ABC News: http://abcnews.go.com/Primetime/story?id=2717038&page=1

Bezos, J. P. (2013). *Biography.* Retrieved January 27, 2013, from http://www.biography.com/people/jeff-bezos-9542209

Black, K. (2002, August 6). Lessons of Hiroshima: Using nuclear bombs on Japan was a political, not military, decision. *Toronto Star.*

Bloomberg Businessweek Magazine website. (2006). *The art of motivation.* Retrieved April 30, 2006, from BloombergBusinessweek: http://www.businessweek.com/stories/2006-04-30/the-art-of-motivation

Bloomberg. (2010). *Southwest Airlines selects spigit to drive innovation.* Retrieved March 1, 2013 from Bloomberg: http://www.bloomberg.com/apps/news?pid=newsarchive&sid=ahHQz0iRecxo

Boches, E. (2012). *The paper clip: A creative exercise*. Retrieved February 15, 2013, from http://edwardboches.com/the-paper-clip-a-creative-exercise

Brainstorming.co.uk. (2006). *How to use the SCAMPER technique*. Retrieved May 9, 2013 from Infinite Innovations Ltd.: http://www.brainstorming.co.uk/tutorials/scampertutorial.html

Butler, C. (2002). *Postmodernism: A very short introduction*. United Kingdom, UK: Oxford University Press.

Buxton, B. (2007). *Sketching user experiences: Getting the design right and the right design*. San Francisco, CA: Morgan Kaufman.

Buzan, T. (1983). *Use both sides of your brain*. US: Dutton.

Byrne, R. (2007). *The Secret*. New York, NY: Atria Books.

Carr, A. (2013). *Nike: The no. 1 most innovative company of 2013*. Retrieved April 10, 2013, from Fast Company's: http://www.fastcompany.com/most-innovative-companies/2013/nike

Cerf, V. G. (2011). *How to fire up U.S. innovation: We need more hands-on tech education for American children, but we also need to keep attracting the best talent from abroad*. Retrieved March 1, 2013, from http://online.wsj.com/article/SB10001424052748704461304576216911954533514.html

Charyton, C., & Snelbecker, G. E. (2007). *General, artistic and scientific creativity attributes of engineering and music students*. Retrieved May 2007, from MIT: http://web.mit.edu/monicaru/Public/old%20stuff/For%20Dava/Grad%20Library.Data/PDF/10400410701397271-1561661185/10400410701397271.pdf

Cherry, K. (n.d.). *Left brain vs. right brain: Understanding the myth and reality of left brain and right brain dominance*. Retrieved April 1, 2013, from the About.com: http://psychology.about.com/od/cognitivepsychology/a/left-brain-right-brain.htm

Chesbrough, H. W. (2003). *Open innovation: The new imperative for creating and profiting from technology*. Boston, MA: Harvard Business School Press.

Collins, J., & Porras, J. I. (1997). *Built to last*. New York, NY: HarperCollins Publisher.

Constable, G., Somerville, B., & Armstron, N. (2003). *A century of innovation: Twenty engineering achievements that transformed our lives*. US: Joseph Henry Press.

Crow, M. (2011). *A New American university reader: Selected writings on university design and related topics*. Retrieved March 1, 2013, from Arizona State University's website: http://president.asu.edu/sites/default/files/New%20American%20University%20Reader%20072611%20(2).pdf

de Bono, E. (1990a). *Lateral thinking: Creativity step-by-step*. New York, NY: HarperCollins.

de Bono, E. (1990b). *The use of lateral thinking* (Ist ed.). United Kingdom, UK: Penguin Books Ltd.
de Bono, E. (1992). *Serious creativity.* New York, NY: Harper Collins.
de Bono, E. (1999). *Six thinking hats.* New York, NY: Little, Brown & Co.
de Bono, E. (2005). *de Bono's thinking course.* New York, NY: Barnes & Noble.
Deschamps, J. P. (2008). *Innovation leaders: How senior executives stimulate, steer and promote innovation* (Ist ed.). San Francisco, CA: Jossey-Bass.
disneyinstitute.com. (n.d.). *Disney's approach to creativity & innovation.* Retrieved April 30, 2013, from http://disneyinstitute.com/courses/creativity_innovation.aspx
Donofrio, N. (2004). *Innovation: The new reality for national prosperity.* 21st century innovation working group recommendations, Version 2.1, 15 December 2004.
Edison, T. A. (2013). *Biography.* Retrieved January 25, 2013, from http://www.biography.com/people/thomas-edison-9284349
Edmunds.com. (2013). *Mazda MX-5 miata history.* Retrieved January 15, 2015, from http://www.edmunds.com/mazda/mx-5-miata/history.html
Einstein, A. (2013). *The biography channel website.* Retrieved Jan 25, 2013, from http://www.biography.com/people/albert-einstein-9285408
Gates, B. (2005, December). The road ahead: How "intelligent agents" and mind-mappers are taking our information democracy to the next stage. *Newsweek.*
Gates, B. (2013). *Biography.* Retrieved January 27, 2013, from http://www.biography.com/people/bill-gates-9307520
Gore, A. (2013). *The future: Six drivers of global change.* New York, NY: Random House.
Hare, B. (2013). *How Marissa Mayer writes her own rules.* Retrieved March 15, 2013, from CNN Tech: http://www.cnn.com/2013/03/12/tech/web/marissa-mayer-yahoo-profile/
Harris, L. V. (2004). *Dancing with dragons: Social construction of technology during times of resource stress.* Ann Arbor, MI: ProQuest.
Harris, L. V. (2005a). *How to think differently: The philosophy of media arts technology.* Honolulu, HI: Hawaii International Conference on Arts and Humanities Conference.
Harris, L. V. (2005b). *Instructor's notes.* Phoenix, AZ: Arizona State University.
Harris, L. V. (2006a). *The IDeaLab needs assessment of today's industry professionals.* Chicago, IL: ASEE (American Society of Engineering Educators) Conference Proceedings.
Harris, L. V. (2006b). *The IDeaLaboratory: Creative and innovative problem solving with technology-based solutions.* Portland, OR: NCIIA (National Collegiate Inventors and Innovators Alliance) Conference Proceedings.

Hayes, M. A., & Comer, M. D. (2010). *Start with humility: Lessons from America's quiet CEOs on how to build trust and inspire followers*. Westfield, IN: The Robert K., Greenleaf Center for Servant Leadership.

Hernandez, J. S., & Varkey, P. (2008). *Vertical versus lateral thinking*. Retrieved April 15, 2013, from http://www.howardluksmd.com/public/Hernandez.pdf

Hindle, T. (2008). *The economist guide to management ideas and gurus*. United Kingdom, UK: Profile Books.

House, C., & Price, R. (2009). *The HP phenomenon: Innovation and business transformation*. Stanford, CA: Stanford University Press.

HP. (n.d.). *IT infrastructure outsourcing services: Balance efficiency and optimization*. Retrieved January 5, 2013, from HP: http://www8.hp.com/us/en/business-services/it-services.html?compURI=1079635

Hymowitz, C., & Murray, M. (1999). *General Electric's Welch discusses his ideas on motivating employees*. Retrieved November 1, 2012, from http://www.wright.edu/~tdung/welch.htm

IDEAlliance conference (ipa.org). (2006). *Branding panel*. Scottsdale, AZ: Gary Springer of Kimberly-Clark Corporation.

Isaacson, W. (2011). *Steve Jobs*. New York, NY: Simon & Schuster.

Ishikawa, K. (1991). *What is total quality control? The Japanese way*. Upper Saddle River, NJ: Prentice-Hall Ltd.

Iverson, K. (1998). *Plain talk: Lessons from a business maverick*. US: John Wiley & Sons, Inc.

Jeffery, J. (2013). *3D printing human organs, but where's the money for It?: Texas startup is on cusp of 3D printing human organs but it must contend with making the technology cost effective first*. Retrieved July 30, 2013 from The Guardian: http://www.theguardian.com/technology/2013/jul/17/3d-printing-organs-money

Kaplan, S. (1997). *Introduction to TRIZ*. Southfield, MI: Ideation International, Inc.

Kelly, T., & Littman, J. (2001). *The art of innovation: Lessons in creativity from IDEO, America's leading design firm*. New York, NY: Currency.

Kim, W. C., & Mauborgne, R. (2005). *Blue ocean strategy: How to create uncontested market space and make competition irrelevant*. Boston, MA: Harvard Business School Publishing.

King, R. (2010). *Workers of the world, innovate*. Retrieved February 15, 2013, from *BloombergBusinessweek:* http://www.businessweek.com/stories/2010-09-08/workers-of-the-world-innovatebusinessweek-business-news-stock-market-and-financial-advice

Lampton, C. (n.d.). *Top 10 Leonardo da Vinci inventions*. Retrieved June 15, 2013, from HowStuffWorks: http://science.howstuffworks.com/innovation/famous-inventors/10-leonardo-da-vinci-inventions.htm#page=10

Leckart, S. (2013). *How 3-D printing body parts will revolutionize medicine: Welcome to the age of bioprinting, where the machines we've built are building bits and pieces of us.* Retrieved August 7, 2013, from Science: http://www.popsci.com/science/article/2013-07/how-3-d-printing-body-parts-will-revolutionize-medicine

Leifer, R. (1998). *An information processing approach to breakthrough innovations facilitating the fuzzy front end of breakthrough innovations* (pp. 130–135). San Juan, PR: IEEE, International Engineering Management Conference (IEMC).

Levy, S. (2000). *Insanely great: The life and times of Macintosh, the computer that changed everything.* New York, NY: Penguin Books.

Li, C., & Solis, B. (2013). *The evolution of social business: Six stages of social business transformation.* Retrieved April 15, 2013, from http://www.slideshare.net/Altimeter/the-evolution-of-social-business-six-stages-of-social-media-transformation

Linstone, H., & Turoff, M. (2002). *The delphi method: Techniques and applications.* Boston, MA: Addison-Wesley Publishing Company, Inc.

Lohr, S. (2009, February 25). In innovation, U.S. said to be losing competitive edge. *The New York Times.*

MacKay, H. (2012, November 26). Career and life lessons. *The Arizona Republic.*

Mariaraj, V. J. (2005). *Boeing, Oracle, EDS, and other high profile companies have drawn great benefits from mind mapping.* Retrieved June 20, 2013, from Ezine: http://ezinearticles.com/?Boeing,-Oracle,-EDS,-And-Other-High-Profile-Companies-Have-Drawn-Great-Benefits-from-Mind-Mapping&id=106728

McKinney, P. (2012). *Beyond the obvious: Killer questions that spark game-changing innovation*, Santa Clara, CA: Hyperion.

Michalko, M. (2011). *Creative thinkering: Putting your imagination to work.* Novato, CA: New World Library.

National Academy of Engineering (2005). *Educating the engineer of 2020: adapting engineering education to the new century.* Washington, DC: National Academies Press.

Nolan, V., & Williams, C. (2010). *Synectics as a creative problem solving (CPS) system.* Boston, MA: Synecticsworld, Inc.

Novack, J. D., & Cañas, A. J. (2006). *Underlying concept maps and how to construct and use them.* Retrieved December 15, 2012, from the Institute for Human and Machine Cognition: http://cmap.ihmc.us/publications/researchpapers/theorycmaps/theoryunderlyingconceptmaps.htm

Osborn, A. F. (1979). *Applied imagination: Principles and procedures of creative problem solving.* New York, NY: Charles Scribner's Sons.

Pasachoff, N. (1996). *Marie Curie and the science of radioactivity.* United Kingdom, UK: Oxford University Press.

Pausch, R., & Zaslow, J. (2008). *The last lecture: Really achieving your childhood dreams.* United Kingdom, UK: Hodder & Stoughton.

Piccioni, R. (2011). *A world without Einstein.* Westlake Village, CA: Real Science Publishing.

Pitney Bowes. (n.d.-a). *Corporate overview.* Retrieved March 1, 2013, from Pitney Bowes: http://news.pb.com/corporate-overview/

Pitney Bowes. (n.d.-b). *Pitney Bowes expands offerings with new color product.* Retrieved March 1, 2013, from Pitney Bowes: http://news.pb.com/press-releases/pitney-bowes-expands-offerings-with-new-color-production-print-system.htm

Pitney Bowes. (n.d.-c). *Super-sized service: Pitney Bowes teams with McDonald's to deliver efficiency, innovation and savings.* Retrieved March 1, 2013, from Pitney Bowes: http://www.pb.com/docs/US/pdf/Microsite/Management-Services/news-and-resources/case-studies/mcdonalds-case-study-2012.pdf

Poeter, D. (2011). *Top PC exec departs HP's personal systems group.* Retrieved March 1, 2013, from *PCmag.com:* http://www.pcmag.com/article2/0,2817,2395647,00.asp

Power, W. D. (2013). *How Nike's innovation kitchen made it Fast Company's most innovative company in 2013.* Retrieved April 11, 2013, from W. Dave Power's: http://wdpower.wordpress.com/2013/04/01/how-nikes-innovation-kitchen-made-it-fast-companys-most-innovative-company-in-2013/

Report to the President (2012). *Engage to excel: Producing one million additional college graduates with degrees in science, technology, engineering, and mathematics.* Retrieved April 1, 2013, from Executive Office of the President, President's Council of advisors on Science and Technology (PCAST) website: http://www.whitehouse.gov/administration/eop/ostp/pcast

Roberts, K. (2011). *The origin of business, money, and markets.* New York, NY: Columbia Business School Publishing.

Roberts, P. (1998, October). John Deere runs on chaos: How one company factory combines two trendsetting ideas: Mass customization and complexity theory. *Fast Company.*

Rother, M. (2010). *Toyota kata: Managing people for improvement, adaptiveness, and superior results.* New York, NY: McGraw-Hill.

Rothman, W. (2009). *The dirty backstabbing mess called betamax vs VHS.* Retrieved March 1, 2013, from GIZMO: http://gizmodo.com/5316722/the-dirty-backstabbing-mess-called-betamax-vs-vhs

Sadowski, M., & Sorby, S. (2012). A delphi study as a first step in developing a content inventory for engineering graphics. *Proceedings of the 2012 Annual Engineering Design Graphics Midyear Meeting,* 126–132.

Schwartz, A. (2009). *Researchers develop super-flexible, rubber-like OLED display.,* Retrieved June 17, 2013, from *FastCompany:* http://www.fastcompany.com/1280228/researchers-develop-super-flexible-rubber-oled-display

Showkeir, M., & Showkeir, J. (2013). *Yoga wisdom at work: Finding sanity off the mat and on the job.* San Francisco, CA: Berrett-Koehler Publishers, Inc.

SHRM (Society for Human Resource Management). (2012). *2012 employee benefits: The employee benefits landscape in a recovering economy.* Alexandria, VA: Colonial Life.

Slaughter, S., & Rhoades, G. (2004). *Academic capitalism and the new economy: markets, state, and higher education.* Baltimore, MD: The John Hopkins University Press.

Sloane, P. (2012). *Divergent and convergent thinking.* Retrieved January 30, 2013, from Innovation Excellence: http://www.innovationexcellence.com/blog/2012/10/24/divergent-and-convergent-thinking/

Smith, S. (2013). *Coming soon to a 3D printer near you: Human tissues and organs.* Retrieved July 29, 2013, from Quartz: http://qz.com/78877/how-soon-will-we-be-able-to-3-d-print-entire-human-organs-sooner-than-you-think/

Stanovich, K. E. (2009). *What intelligence tests miss: The psychology of rational thought.* New Haven, CT: Yale University Press.

TED. (2012). *Riveting talks by remarkable people, free to the world.* Retrieved September 18, 2012, from TED: Ideas worth spreading: www.ted.com

Terniko, J., Zusman, A., & Zlotin, B. (1997). *Systematic innovation: An introduction to TRIZ (Theory of Inventive Problem Solving).* United Kingdom, UK: CRC Press.

The Bridge to the Future website. (2012). *The board.* Retrieved May 1, 2013, from San Francisco BASOC Olympic: http://www.basoc.org/basoc2012/board_harris_p.html

The Incentive Federation, Inc. (n.d.). *Incentive legislation campaign frequently asked questions.* Retrieved July 1, 2013, from http://incentivefederation.org/displaycommon.cfm?an=1&subarticlenbr=8

Vishwanath, S. (2013). *The power of visualization: Meditation secrets that matter the most.* Mumbai Area, India: Soul Power Magic Publisher.

Walters, H. (2008). *Apple's design process.* Retrieved May 23, 2013, from *Bloomberg BusinessWeek:* http://www.businessweek.com/the_thread/techbeat/archives/2008/03/apples_design_p.html

Zuckerberg, M. (2013). *Biography.* Retrieved January 27, 2013, from http://www.biography.com/people/mark-zuckerberg-507402

# Index

## A
*Academic Capitalism and the New Economy: Markets, State, and Higher Education* (Slaughter and Rhoades), 164
AFV tool, 130
Amazon.com, 27
Apple Corporation, 30
Architecture of, idea, 135
*The Arizona Republic* (Maren and Pausch), 14

## B
Bath, Patricia, 25–27
Bezos, Jeff, 27–28
Blame Game, 6–7
Brainstorming, 102–106
   examples of professional, 105–106
   limitations, 104–105
   rules of, 103–104
   using lateral thinking, 109–110
Brainwriting, 111–112
Business evolution, 61–62
Business models of innovation
   ideas + leaders, 63–64
   ideas + leaders + teams + plans, 68–70
   ideas + motivation, 67
   ideas + processes, 64–66
Butts, Lucifer Gorgonzola, 19
Buy-in and mutual benefits, 12–13

## C
Calming your mind, 14
*A Century of Innovation: Twenty Engineering Achievements That Transformed Our Lives* (Constable, Somerville, and Armstrong)
Chemical and civil engineers, 151
Chill out, 8
Clear problem statement, 90–91
Collaborative best practices team, 77–78
Common sense, 2–3
Communication causes, problem, 87
Convergent *vs.* divergent thinking, 96–97
Creative thinking
   brainstorming, 102–106
   brainwriting, 111–112
   convergent *vs.* divergent thinking, 96–97
   *vs.* critical thinking, 35–37
   idea conceptualization phase, 101
   inspiration, 97
   inspiring your idea team, 98–101
   lateral thinking, 107–111
   metaphoric thinking, 112–114
   SCAMPER, 114–115
   synectics, 115–118
   TRIZ, 118–119
Critical thinking *vs.* creative thinking, 35–37
Cross-pollinating expert team/project team, 73–74
Cross-pollinating senior management team, 72–73
Curie, Marie, 20–21

## D
DARPA. *See* Defense Advanced Research Projects Agency
da Vinci, Leonardo, 24–25
Deere & Company, 72
Deere, John, 65, 72
Defense Advanced Research Projects Agency (DARPA), 40
Delphi tool, 127–128
Description drawing, 141–142

Designing and testing
  hard work, 133–134
  mock-ups, 142–148
  project management, 134–136
  prototypes, 148–152
  renderings, 136–142
Divergent thinking *vs.* convergent, 96–97
3D printing technology prototypes, 150

**E**
Economic factors
  academic capitalism, 164–166
  drivers and copycats, 162–163
  VHS *vs.* Betamax, 163–164
Economy and innovation, 39–40
Edison, Thomas Alva, 21–22
Education, 40–41
Einstein, Albert, 22–24
Electronic engineering physical prototype models, 149
Environmental causes, problem, 87

**F**
Facebook, Inc., 27
Failure, 6–7
Fear of external global partners, 46–48
Fear of failure, 51–52
Financial literacy, 6
"FuelBand,", 69

**G**
Gates, Bill, 28–30
General Electric, 52
Go/No-Go decision, 74, 75
Gordon, William J. J., 115
Group consensus, idea criteria tool, 125–127

**H**
Harris, Peter L., 63
Hewlett Packard's (HP's) Innovation Program Office, 46
High- and low-impact innovation, 167–168
Human capital causes, problem, 87

**I**
Idea conceptualization phase, 101
*Idea Engineering: Creative Thinking and Innovation* (Pausch), 2
Idea engineering, future of, 170–172
Idea engineers. *See also* Modern-day visionaries
  Curie, Marie, 20–21
  da Vinci, Leonardo, 24–25
  Edison, Thomas Alva, 21–22
  Einstein, Albert, 22–24
Idea selection preliminary work
  problem and ideas, 121–122
  selection criteria, 122–124
  user profile, 122
Idea teams
  collaborative best practices team, 77–78
  cross-pollinating expert team/project team, 73–74
  cross-pollinating senior management team, 72–73
  managed partnership, 79
  open innovation team, 75–77
  senior enterprise team, 74–75
  thought leaders network team, 70–72
  and workforce teams, 80–82
"Imagination's Playground," 47
Individual ratings, idea criteria tool, 127
Information Technology and Innovation Foundation (ITIF), 39
Innovation, 39–40. *See also* Models of innovation
  high- and low-impact, 167–168
  models of, 61–82
Innovators *vs.* visionaries, 17–19
Inspiration, 97
Inspiring your idea team, 98–101
Intelligence, 2–3
Internet marketing concept map, 89
Ishikawa's fishbone diagram, 88
ITIF. *See* Information Technology and Innovation Foundation
Iverson, F. Kenneth, 67

## J

*Jimmy Kimmel Live,* (television venues), 20
Jobs, Steven Paul, 30–32, 64
Johnson, R. W. Jr., 51

## L

Lack of resources
  human capital, 48–49
  money, 49
  project space, 48
  time, 48
Lack of visionary leadership, 44–46
Laserphaco Probe, 25
*The Last Lecture: Really Achieving Your Childhood Dreams* (Pausch), 2
Lateral thinking, 107–111
Live a life, 1–2
Live in here and now, 8

## M

Managed partnership, 79
Mayer, Marissa, 44
Mechanical engineering physical prototype models, 149
Memory drawing, 137–139
Metaphoric thinking, 112–114
Microsoft, 28
Mock-ups, 142–148
Models of innovation
  business evolution, 61–62
  business models, 62–70
  dedicated idea team, 78–80
  idea teams and workforce teams, 80–82
  seven idea team models, 70–78
Modern-day visionaries
  Bath, Patricia, 25–27
  Bezos, Jeff, 27–28
  Gates, Bill, 28–30
  Jobs, Steven Paul, 30–32
  Zuckerberg, Mark, 27
Music meditation, 15

## N

National Football League (NFL), 63
Need-based innovation, 37
*Newton's Apple,* (television venues), 20

NeXT Computer, 32
NFL. *See* National Football League

## O

ODMs. *See* Outsourced design manufacturers
Old Asian cliché, 3
"Open innovation," 76
Open innovation team, 75–77
Optiflex software, 65
Organizational roadblocks
  fear of external global partners, 46–48
  fear of failure, 51–52
  lack of resources, 48–49
  lack of visionary leadership, 44–46
  poor internal communication, 49–51
  resistance to change, 53–54
  reward system, 52–53
Osborn, Alex Faickney, 102
Outsourced design manufacturers (ODMs), 47

## P

Parker, Mark, 68
PCT. *See* Purdue creativity test
People resist change, 54–60
Perfection, 6–7
*The Phoenix Gazette,* 38
Political factors
  atom bomb, 167
  long island overpass design, 166–167
Poor internal communication, 49–51
Presentation drawing, 139–140
Prince, George M., 115
Priority tool, 128–129
Problem identification
  clear problem statement, 90–91
  concept and mind mapping, 88–90
  deconstructing, 84–87
  of defining a problem, 91–94
  fishbone diagram, 87–88
Process causes, problem, 87
Project management, 134–136
Project Team, 73

Pros & Cons tool, 130
Purdue creativity test (PCT), 99

R
Reality, perceptions of, 8–10
Relationship model, 10–12
Renderings
　description drawing, 141–142
　memory drawing, 137–139
　presentation drawing, 139–140
　sketch, 136–137
　technical drawing, 140–141
Repeated words of affirmation, 15–16
Resistance to change, 53–54. *See also* People resist change
Resource causes, problem, 86
Respect, 5
Reward system, 52–53
"Rube Goldberg," defined, 19–20

S
"Sandbox" approach, 71
SCAMPER, 114–115
Screening tools
　AFV tool, 130
　Delphi tool, 127–128
　group consensus, idea criteria tool, 125–127
　individual ratings, idea criteria tool, 127
　priority tool, 128–129
　Pros & Cons tool, 130
　voting tool, 129
Senior enterprise team, 74–75
Serendipity and defining moments, 172–174
*Six Thinking Hats* (de Bono, Edward), 50
Sketch, 136–137
*Sketching User Experiences: Getting the design right and the right design* (Buxton, Bill), 136
Social factors, 157–161
　single-serve coffee machines, 159–160
　videotex technology, 160–161
Standard IQ tests, 3

Stinkin' thinkin', 3–5
Synectics, 115–118

T
Technical drawing, 140–141
Testing, 152–154. *See also* Designing and testing
Thought leaders network team, 70–72
*The Today Show,* (television venues), 20
*The Tonight Show,* (television venues), 20
Training and education, 40–41
TRIZ, 118–119

U
User experience (UX) design, 151–152
UX design. *See* User experience design

V
Visionaries. *See also* Modern-day visionaries
　*vs.* innovators, 17–19
Vision boards, 15
Visualization, 14–15
Voting tool, 129

W
Walt Disney, 63, 64
Welsh, Jack, 52
*What Intelligence Tests Miss,* (Stanovich, Keith), 2
*Whispered Words: Your Story, Your Culture,* 77
Workforce teams and idea teams, 80–82

Y
*Yoga Wisdom at Work: Finding Sanity Off the Mat and On the Job* (Maren and Showkeir), 14
You are not superman (or superwoman), 5–6

Z
Zuckerberg, Mark, 27

## THIS BOOK IS IN OUR INDUSTRIAL, SYSTEMS, AND INNOVATION ENGINEERING COLLECTION

William R. Peterson, Editor

Momentum Press is dedicated to developing collections of complementary titles within specific engineering disciplines and across topics of interest. Each collection is led by a collection editor or editors who actively chart the strategic direction of the collection, assist authors in focusing the work in a concise and applied direction, and help deliver immediately actionable concepts for advanced engineering students for course reading and reference.

Some of our collections include:

- *Manufacturing and Processes*—Wayne Hung, Editor
- *Manufacturing Design*
- *Engineering Management*—Carl Chang, Editor
- *Electrical Power*
- *Communications and Signal Processing*—Orlando Baiocchi, Editor
- *Electronic Circuits and Semiconductor Devices*—Ashok Goel, Editor
- *Integrated Circuit Design*
- *Sensors, Control Systems and Signal Processing*
- *Antennas, Waveguides and Propagation*
- *Thermal Engineering*—Derek Dunn-Rankin, Editor
- *Fluid Mechanics*—Dr. George D. Catalano, Editor
- *Environmental Engineering*—Francis Hopcroft, Editor
- *Geotechnical Engineering*—Dr. Hiroshan Hettiarachchi, Editor
- *Transportation Engineering*
- *Sustainable Systems Engineering*—Dr. Mohammad Noori, Editor
- *Structural Engineering*
- *Chemical Reaction Engineering*
- *Chemical Plant & Process Design*
- *Thermal and Kinetics Topics in Chemical Engineering*
- *Petroleum Engineering*
- *Materials Characterization and Analysis*—Dr. Richard Brundle, Editor
- *Mechanics & Properties of Materials*
- *Computational Materials Science*
- *Biomaterials*

Momentum Press is actively seeking collection editors and authors. For more information about becoming an MP author or collection editor, please visit **http://www.momentumpress.net/contact** and let us hear from you.

---

### Announcing Digital Content Crafted by Librarians

Momentum Press offers digital content as authoritative treatments of advanced engineering topics, by leaders in their fields. Hosted on ebrary, MP provides practitioners, researchers, faculty and students in engineering, science and industry with innovative electronic content in sensors and controls engineering, advanced energy engineering, manufacturing, and materials science. **Momentum Press offers library-friendly terms:**

- perpetual access for a one-time fee
- no subscriptions or access fees required
- unlimited concurrent usage permitted
- downloadable PDFs provided
- free MARC records included
- free trials

The **Momentum Press** digital library is very affordable, with no obligation to buy in future years.

For more information, please visit **www.momentumpress.net/library** or to set up a trial in the US, please contact **mpsales@globalepress.com**.

T 49.5 .H275 2014
Harris, La Verne Abe,
Idea engineering
10/2015